Intelligent Transportation Systems

Intelligent Transportation Systems

Editor

Teodor Piatek

scitus
academics

Intelligent Transportation Systems
Edited by **Teodor Piatek**

ISBN: 978-1-68117-539-3
Library of Congress Control Number: 2016942939

SCITUS Academics LLC,
www.scitusacademics.com
Box No. 4766, 616 Corporate Way,
Suite 2, Valley Cottage,
NY 10989

Notice

Printed in United States of America on Acid Free Paper

Preface

Intelligent transport systems (ITS) are essential for improving safety and addressing the growing problems of emissions and congestion. They could make transport safer, more efficient and more sustainable by using a variety of communication and information technologies in other modes of transport for passengers and freight. The integration of existing technologies can additionally create new services. It is the key to supporting growth and job creation in the transport sector. Intelligent transport systems are systems that use synergistic systems and technology concepts to develop and improve all types of transport systems. Intelligent transport systems (ITS) are advanced applications that aim to provide innovative services relating to different modes of transport and traffic management, enabling different people to be better informed and to make transport networks safer, more coordinated and more intelligent. Intelligent transport systems vary in technology, from simple management systems, including automotive navigation; traffic signal control systems; container management systems; variable message signs; automatic number plate recognition or possibly speed cameras for monitoring applications, such as CCTV security systems; and innovative applications that integrate live feedback and data. In addition, predictive techniques for advanced modeling and comparison with historical baseline data are being developed.

This book focuses on various ITS technologies and applications and management practices in the hope that the open exchange of scientific results and ideas presented in this book will lead to a better understanding of ITS technologies and their applications. The scope of this book covers the promotion, consolidation and coordination of ITS technical activities and focuses both internally and externally on cooperative activities.

Table of Contents

Chapter 1

Detection of Emergency Telephone Indicators in a Tunnel Environment

Zhipeng Wang[1] , Matasaka Kagesawa[1], Shintaro Ono[1], Atsuhiko Banno[2], Takeshi Oishi[1], Katsushi Ikeuchi[1]

[1]Institute of Industrial Science, The University of Tokyo, Tokyo, Japan

[2]Advanced Industrial Science and Technology, Ibaraki, Japan

ABSTRACT

Positioning of vehicles is important for ITS. In a tunnel environment, most positioning solutions based on GPS sensors or ordinary cameras will fail. For positioning we propose a method to detect emergency telephone indicators in a tunnel environment by using infrared cameras. The proposed detection method makes use of both appearance and motion information of the target objects. By optimizing the detecting pipeline, the method works in real time and produced 100 % detection rate and 0 % false alarm rate in one of our experiments.

KEYWORDS

Object detection; Positioning

1 INTRODUCTION

Positioning of vehicles acts as a fundamental role in autonomous driving, and is also of great importance for driving assistance, vehicle navigation, etc. When GPS sensors function properly, the task is easy. While in a tunnel environment, there are no GPS signals available for most of the time. A new positioning system which functions properly in a tunnel environment is necessary [6]. In this paper we propose an object detection method which can be used for positioning systems in tunnels.

Our proposal is part of an automated driving system in a NEDO project, “Development of Energy-saving ITS Technologies”. The automated driving system is vehicle-oriented, and an express way is the main application scenario. No specific facilities are assumed to exist on road sides, while instead, the experimental vehicles are equipped with sensors and vehicle-to-vehicle communication systems. There are a few sensors used for positioning. For example, sensors used to extract white road lines to estimate the lateral position in the lane, GPS sensors, dead reckoning systems, and stereo far-infrared camera systems intended for obstacle detection. On the street, GPS sensors can be used for positioning. While positioning in tunnels is difficult since GPS signals are not available and no specific equipment on road sides is assumed. For positioning in tunnels, GPS sensors are used to record the position of tunnel entrances, and dead reckoning systems are used to infer position by continuously sensing the vehicles’ speed and direction. However, errors of dead reckoning systems will accumulate. Thus, the proposed method uses far-infrared cameras installed on the vehicles to detect signs in the tunnels, which contain position information, and will be used to eliminate the accumulated errors.

In most tunnels on the expressways in Japan, there are many signs appearing at equal intervals. We focus on the emergency telephone indicators, which appear every 200 meters in tunnels. The absolute coordinates of the emergency telephone indicators can be obtained by the method of [30]. If the emergency telephone indicators can be sensed while traveling in tunnels, and the distance from the vehicle to the indicators can be estimated, then this information can be used to eliminate accumulated errors of dead reckoning systems. Detection methods, e.g., [29], based on ordinary cameras fail due to darkness. Here we use far-infrared cameras, which are suitable in dark environments and are already installed on our experimental vehicles. Inspired by a previous work [28], we propose an approach to detect emergency telephone indicators.

Detection performance and efficiency are the two important aspects of our method. In a tunnel environment, in addition to the target objects, a lot of noisy objects also appear, e.g. ordinary lights, other vehicles, and other vehicles' shadows. And some of the noisy objects cannot even be distinguished from the target objects by appearance, as shown in Figure 1. The clutter property of the sensed data makes the detection challenging. Our method meets this challenge by making use of both appearance and temporal information of the target objects. There are two main steps in the method. The first step deals with keypoints. It takes original data as input, and outputs keypoint clusters as detection hypotheses. In this step, keypoints are detected, verified and then clustered. To detect keypoints, all points on each frame are uniformly sampled and filtered with pre-set intensity thresholds. Then the keypoints are verified by a simple keypoint appearance model built by *k*-means. At the end of the first step, the keypoints are clustered based on the Euclidean distance. The second step takes the keypoint clusters as input, verifies them by appearance and temporal information, and outputs the keypoints which pass verification, as detection results. In the second step, the keypoint clusters are labeled based on appearance by an Adaboost machine, which is trained using intensity histograms of keypoint clusters from target objects and keypoint clusters from noisy objects. The keypoint clusters are also tracked by temporal association through frames. Motion information encoded in the trajectories are used to further verify the keypoint clusters. Finally, the keypoint clusters which pass both appearance and temporal verifications are confirmed as emergency telephone indicators.

This pipeline is also designed with consideration for the requirement of efficiency. The method deals with the large amounts of information contained on one frame, following a hierarchical manner. The later a step is, the more time-consuming it is, and the fewer instances it deals with. From an image containing 10^5 pixels, 10^4 points go through the keypoint detection step of testing by intensity thresholds. Then in average, 10^3keypoints are detected, and verified, leaving about 10^2 keypoints to be clustered. Afterwards, fewer than 10 keypoint clusters are left; these are dealt with by the very time-consuming steps of generating image features and tracking.

The advantage of our method is its ability to give promising detection results from cluttered data in real time. In addition, this method successfully combines bottom-up and classification methods, as well as combines both appearance and temporal information.

Figure 1. Original data and detection results. In (**a**), the *red arrow* points to the target object: emergency telephone indicator, and the *green arrows* point to noisy objects. In (**b**), *red rectangles* mark detection hypotheses labeled as positive using appearance information, and *green rectangles* mark negative ones. *Yellow* trajectories mark detection hypotheses labeled as positive using temporal information, and *white* trajectories mark negative ones

The paper is organized as follows: Section 2 reviews related work, Section 3proposes a pipeline for the method, Section 4 gives experimental results, and Section 5 concludes.

2 RELATED WORK

Most modern detection methods fall into two categories. Some [5, 8, 10, 14, 19, 25, 27,32] follow the sliding-window schema, and they detect objects by considering whether each of the sub-images contains an instance of the target object. Classifiers are usually employed by these methods. The other methods [1, 7, 9, 16, 20–22] infer object centers based on local image features in a bottom-up manner. The proposed method takes advantages of both frameworks. Following the bottom-up manner, keypoints are detected, verified, and clustered. After these steps, the keypoint clusters are considered as detection hypotheses. Then following the sliding-window schema, the keypoint clusters are verified by their appearance and temporal information, using discriminative methods. Previous methods [24] also consider the combination of the two frameworks. Detection hypotheses are gained using a Hough transform and then verified by support vector machines in [20, 31]. The methods in [11, 23], use randomized decision trees for both decisions: whether local features belong to foreground objects or not, and decisions of their Hough votes. The method proposed in [15] describes both frameworks in the same manner. While giving state-of-the-art detection performance, these other methods can't meet the requirement for efficiency as our method does. Our work is also related to feature grouping methods [31], detecting methods using trajectories [3, 4], tracking methods [12, 17], and methods integrating appearance and temporal information [29]. Especially, compared with the method proposed in [28], our method employs a more effective classifying machine by setting biased weights for positive and negative training examples, and far over-performs [28].

3 EMERGENCY TELEPHONE INDICATOR DETECTION

Our proposal can be considered a two-step method. The first step deals with keypoints. It takes original data as input, and outputs keypoint clusters as detection hypotheses. The second step takes these keypoint clusters as input, verifies them by their appearance and motion information, and outputs the ones which pass verifications as detection results.

3.1 Keypoint Detection

In data collected using ordinary cameras, keypoints [2, 18] invariant to rotations, affine changes, and illumination changes are preferable. In our case, keypoint detection is intended to provide hypotheses for emergency telephone indicators. Thus intensity is of great importance. Our method employs a simple yet useful method to detect keypoints. Firstly, points are uniformly sampled for an offset of 6 in width, and 7 in height (the length of an emergency telephone indicator is larger than its width). In this manner the magnitude of instances is reduced by nearly two orders. Then points that pass the test, which verifies them by setting intensity thresholds, are considered as keypoints. Here a Gaussian distribution is assumed for the intensities of the points.

Let $\{\mathbf{X}\Diamond$ denote all the sampled points, I_x the intensity of each point, and I_x the label. If the point is considered as belonging to emergency telephone indicators, $I_x = 1$, otherwise, $I_x = 0$. By setting lower threshold,

I^{th1}_{x}, and higher threshold, I^{th2}_{x}, the probability that points belongs to emergency telephone indicators based on their falling into this interval is given by,

$$P\left(l_{\mathbf{x}} = 1 \mid I_{\mathbf{x}}^{th1} \leq I_x \leq I_{\mathbf{x}}^{th2}\right) = \frac{P\left(l_{\mathbf{x}} = 1, I_{\mathbf{x}}^{th1} \leq I_x \leq I_{\mathbf{x}}^{th2}\right)}{P\left(I_{\mathbf{x}}^{th1} \leq I_x \leq I_{\mathbf{x}}^{th2}\right)} \quad (1)$$

At this step, the possibility that as few points as possible, belonging to the emergency telephone indicators, are excluded, is also considered. The probability of one point falling into the defined interval based on its belonging to emergency telephone indicators is given by,

$$P\left(I_{\mathbf{x}}^{th1} \leq I_x \leq I_{\mathbf{x}}^{th2} \mid l_{\mathbf{x}} = 1\right) = \frac{P\left(l_{\mathbf{x}} = 1, I_{\mathbf{x}}^{th1} \leq I_x \leq I_{\mathbf{x}}^{th2}\right)}{P\left(l_{\mathbf{x}} = 1\right)} \quad (2)$$

Points for which the intensities fall in the pre-set thresholds, are detected as keypoints.

3.2 Keypoint Verification

As shown in Figure 2b, the detected keypoints don't just belong to emergency telephone indicators, but also belong to the background. For training purposes, keypoints belonging to emergency telephone indicators are considered positive, all others are negative.

a b

Figure 2. Keypoint detection

To verify the keypoints, the appearance of the sub-image around each keypoint is used. Intensity histograms are used to describe the appearance. Noisy keypoints not only come from the wall of the tunnel, but also from ordinary lights, other vehicles, and other vehicles' shadows. Thus robust linear classifiers are not suitable for the verification. Here, a general model in the form of a simple mixture is used. The *k*-means method is used to cluster the intensity histograms, $\{ A_x, l_x = 1 \}$, of the positive keypoints, and, $\{ A_x, l_x = 0 \}$ of the negative keypoints.

Let $\{C_1^i, i = 1, 2, \ldots, n_1\}$ denote the intensity histogram centers of the positive keypoints, and $\{C_0^i, i = 1, 2, \ldots, n_2\}$ the negative. For each C_1, the average Euclidean distance between $\{C_0^i, i = 1, 2, \ldots, n_2\}$ is calculated as,

$$Eu\left(C_1^i\right) = \frac{1}{n_2} \sum_{j=1}^{n_2} Euclid\left(C_1^i, C_0^j\right) \qquad (3)$$

Here, $Euclid$ calculates the Euclidean distance, and Eu is an evaluation function of the positive feature centers. The positive feature centers are ranked by Eu and the 10 positive feature centers with the largest Eu are are chosen and used for verification.

For verification, the intensity histogram of each keypoint's surrounding sub-image is extracted. Then the Euclidean distance between the extracted intensity histogram and its nearest positive feature center is calculated. If this distance exceeds a threshold, D^{thAx}, it is considered negative, otherwise it is considered positive. Here, for simplicity, unlike [26], the same threshold is used for all components of the mixture.

3.3 Keypoint Clustering

After the keypoint verification step, on some frames the result is pretty good, while on other frames appearance of the keypoints is not enough to decide whether the keypoints belong to the emergency telephone indicators or not. Here generation of keypoint trajectories is not feasible, since nearby keypoints are similar in appearance and the time complexity of associating such a large number of keypoints along the time dimension is high. So the keypoints are clustered, then data association in time dimension only is needed to deal with a small number of keypoint clusters.

To cluster the keypoints, a minimum spanning tree (mst) is built using the pairwise Euclidean distance between two keypoints. Then the mst is split by cutting edges larger than a threshold. This results in a grouping of the keypoints, denoted by, $\gamma = \{\mathbf{g}\diamond$.

3.4 Keypoint Cluster Verification by Appearance

For each keypoint cluster, the smallest bounding rectangle is considered a detection hypothesis, as shown in Figs. 3c and 3d. There are three main sources of noise: ordinary lights, other vehicles, and other vehicles' shadows. The global appearance of ordinary lights is different from that of the emergency telephone indicators'. As ordinary lights get further from the infrared camera, the intensities of their corresponding sub-images in the collected data gets lower. At a certain distance, the intensities of the ordinary lights are almost the same as the intensities of the emergency telephone indicators'. For ordinary lights of which the intensities are higher than the intensities of emergency telephone indicators', the transition regions from the lights to tunnel walls will have similar intensities as the emergency telephone indicators'. This indicates that although locally the emergency telephone indicators share the same appearance with ordinary lights, globally they can still be distinguished by appearance. As for other vehicles and their shadows, their intensity range is very close to the intensity range of the emergency telephone indicators', and they can hardly be distinguished by appearance alone.

At this step, the keypoint clusters are verified by their appearance, ideally excluding keypoint clusters belonging to ordinary lights. An Adaboost machine is trained using intensity histograms of the emergency telephone indicators and ordinary lights. The appearance of other vehicles is close to that of the emergency telephone indicators, and they are not used for training the machine. For training of the machine, labeled 32-dimensional intensity histograms are firstly normalized. Then each weak classifier of the machine makes a decision on one dimension of the intensity histograms. After this step, each keypoint cluster is either labeled as positive or negative.

In this step, to emphasize the Adaboost machine's performance on the positive training examples, we set the initial weights of the positive training examples 7 times as large as the weights of the negative training examples. Since in practice, whether each keypoint cluster is a target object or not is decided by both appearance and motion information. The difficulties of excluding noisy objects can be left for later steps.

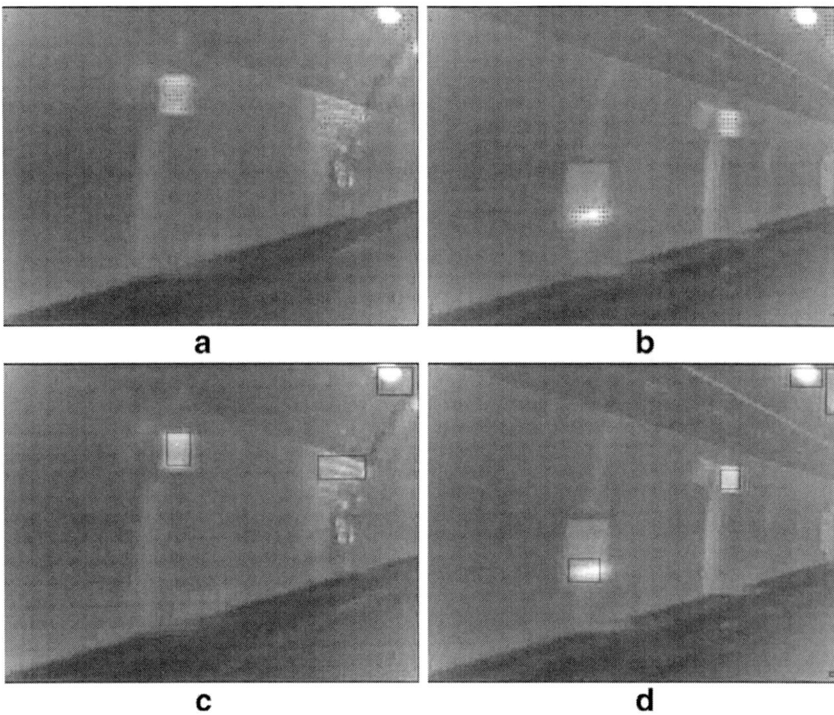

Figure 3. Keypoint verification and clustering. *Red circles* mark keypoints which pass the verification, while *blue marks*failed ones. *Rectangles mark* keypoint clustering results

3.5 Keypoint Cluster Tracking

Not all noisy detection hypotheses can be excluded by using appearance, as shown in Figure 4. To distinguish keypoint clusters belonging to other vehicles and their shadows, the keypoint clusters are tracked through frames to generate trajectories.

In our case of keypoint cluster tracking, the problem is relatively simple, since no occlusion occurs. To keep the method on-line and maintain efficiency, a pool of trajectories are kept, $\tau = \left\{T_g^i, i = 1, 2, \ldots, n\right\}$, and new detection hypotheses act as detection responses, v= $\left\{ n_g^i, i = 1, 2, \ldots, m \right\}$, in tracking. The problem of tracking is modeled by finding the best data association hypothesis, H^*, between the trajectory set and detection response set as,

$$
\begin{aligned}
H^* &= \arg\max_{H \in \eta} P(H \mid \nu) \\
&= \arg\max_{H \in \eta} \left(\prod_{\left(T_g^i, n_g^j\right) \in H} P_{link}\left(n_g^j \mid T_g^i\right) \right)
\end{aligned} \tag{4}
$$

Figure 4. Keypoint cluster verification by appearance. *Red rectangles*: positive detection hypotheses, and *green*: negative detection hypotheses

Let u_{ij} = 1 or 0 indicate if n^j_g is linked to T^i_g or not, and assuming each trajectory can link once, and each detection response can only be linked once, the problem can be modeled as,

$$
\begin{aligned}
&\arg\max_{u_{ij}} \sum_{i=1}^{n} \sum_{j=1}^{m} u_{ij} \ln P_{link}\left(n_g^j \mid T_g^i\right) \\
&s.t.: \ u_{ij} = 0 \text{ or } u_{ij} = 1, \forall i, \forall j; \\
&\qquad \sum_{i=1}^{n} u_{ij} \le 1 \ ; \sum_{j=1}^{m} u_{ij} \le 1 .
\end{aligned}
$$

Here, $P_{link}\left(n_g^j, T_g^i\right)$ is defined by the appearance difference, the scale difference, and the time gap between the last detection response contained in TigTgi and njgngj. While the Hungarian algorithm [13] gives a near-optimal solution, we follow a very simple manner for the solution by finding the best matched pairs and excluding them until no matching pairs can be found.

3.6 Keypoint Cluster Verification by Motion

As shown in Figure 5, the trajectories from keypoint clusters belonging to emergency telephone indicators are different from other objects' trajectories. In this step, the temporal information encoded in the trajectories is used to further verify the keypoint clusters. A linear model is used to fit each trajectory, and the Pearson Correlation Coefficient (PCC) of the fitting is the criteria for the decision. Let $\left(x_g^i, y_g^i\right)$ denote the coordinate of the *i*th element belonging to a trajectory. The linear assumption is that $y_g^i = a_0 + a_1 x_g^i$. The PCC of the fitting is defined as,

$$r = \left| \frac{\sum_i \left(x_g^i - \bar{x}_g\right)\left(y_g^i - \bar{y}_g\right)}{\left[\sum_i \left(x_g^i - \bar{x}_g\right)^2 \cdot \sum_i \left(y_g^i - \bar{y}_g\right)^2\right]^{1/2}} \right| \quad (5)$$

Where *r* is used to decide if the trajectories of the keypoint clusters belong to emergency telephone indicators or not.

3.7 Object Detection

For each keypoint cluster on the current frame, there exists a label given by the Adaboost machine according to its appearance, and the likelihood of fitting its trajectory to a straight line. For each keypoint cluster, it is considered an emergency telephone indicator if and only if its label - given by the Adaboost machine - is positive, its trajectory is longer than l^{th}, and the likelihood of fitting its trajectory to a straight line is larger than r^{th}.

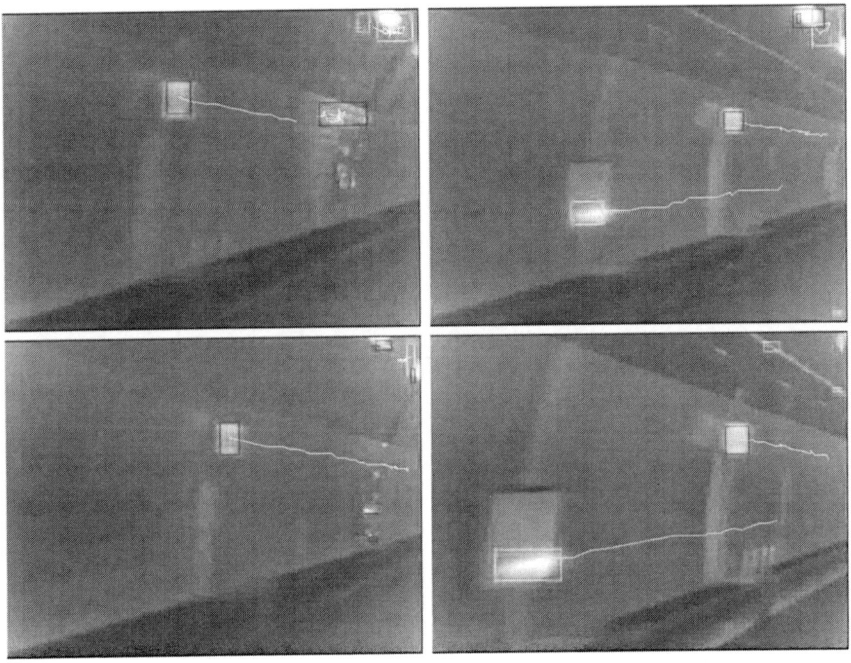

Figgure 5. Detection results

Each trajectory not only connects the detection responses, but also connects the decisions for detection responses made by their appearance and motion patterns. The target objects and noisy objects actually appear in successive frames, and even if we make a wrong decision on one frame, we can expect to recover from this mistake based on the results of other frames. The final results are based on the trajectories of decisions. When one trajectory ends, if more than 80 % of the decisions it connects are positive, then this trajectory is considered positive.

4 EXPERIMENTAL RESULTS

We evaluate our method based on detection performance and efficiency.

4.1 Data

To collect data, we mount an infrared camera on top of the experimental vehicle, and then take several tours of the Awagatake tunnel. Approximately 7,000 frames are collected for each tour. The frame size is 640 × 480, the intensity range is [0,255], and the frame rate is 30 frames per second.

4.2 Implementation Settings

All models are trained using data from the same tour, while evaluated using data from another tour.

To set intensity thresholds for keypoint detection, Gaussian distribution is assumed for the points belonging to emergency telephone indicators. Following the 3σ principle, I^{th1}_{x} is set to 160 and I^{th2}_{x} to 190. The approximate sub-images of the emergency telephone indicators are manually marked, and used for training the mixture model of keypoint verification. The detected keypoints falling into the sub-image are marked as positive, all other points are marked negative. Note that this model and the training method may not be very accurate, since more accurate marking requires more manual effort. About 30,000 intensity histograms of the positive keypoints are sampled, and about 3,000,000 of the negative. When using of *k*-means for clustering the positive intensity histograms, *k* is set to 40; it is set to 400 for the negative. By using these *k* values, both feature sets are over-segmented. The threshold to verify keypoints D^{th}_{Ax} is set to 0.14 for the normalized histograms. For keypoint clustering, the threshold to split the mst is set to 40, which is half the largest height of the emergency telephone indicators. The Adaboost machine used to distinguish other vehicles and their shadows is trained by intensity histograms of positive keypoint clusters and negative keypoint clusters. We manually mark positive and negative keypoint clusters. If the Adaboost machine is trained by averagely weighted training examples, its correct rate on the training examples is overall 84 %. When trained using our bias weighted training examples, its correct rate is 94 % for the positive training examples, and 77 % for the negative training examples. During keypoint cluster tracking, whether a detection response can be linked to a trajectory or not is constrained by position and scale changes. Here scale change limit is set to 4. When the trajectories are fitted as lines, the linear model is also used in associating new detection responses.

4.3 Detection Results

Using an ordinary desktop computer with Intel Core2 Quad 2.6GHz processors, the method deals with real data at a frame rate of 41 frames per second, and this fulfills real-time requirements.

The detection rate and false alarm rate are evaluated on the keypoint clusters as shown in Table 1. More detection results are shown in Figure 5.

Table 1. Detection rate and false alarm rate

Total number	472
Correctly labeled	468
Miss detections	4
False alarms	22
Detection rate	99.2 %
False alarm rate	0.7 %

While evaluated on a much smaller dataset, the detection rate and false alarm rate of [28] are 90 % and 19 % respectively. Our experiment outperforms [28], because our sensed images are much clearer, and also because of our more effective training of the Adaboost machine.

The results on the trajectories of decisions are also evaluated. The method correctly detects all of the 22 emergency telephone indicators with no false alarms. The detection rate is 100 %, and the false alarm rate is 0 %.

5 CONCLUSION

We propose an object detection method to detect emergency telephone indicators in a tunnel environment. The method makes use of appearance and motion information of the target objects in a hierarchical manner. With careful optimization of the detection pipeline, the method gives promising results in real time. Based on the detection results, a positioning system in a tunnel environment can be expected.

ACKNOWLEDGMENTS

The work is in part supported by NEDO. The first author is sponsored by China Scholarship Council. The authors thank Patricia Knapp and Kent Fujiwara for revising suggestions on the manuscript.

REFERENCES

1. Barinova, O., Lempitsky, V., Kohli, P.: On detection of multiple object instances using hough transforms. In: CVPR, pp. 2233–2240 (2010)

2. Bay, H., Tuytelaars, T., Van Gool, L.: Surf: speeded up robust features. In: ECCV, pp. 404–417 (2006)

3. Brostow, G., Cipolla, R.: Unsupervised bayesian detection of independent motion in crowds. In: CVPR, vol. I, pp. 594–601 (2006)

4. Brox, T., Malik, J.: Object segmentation by long term analysis of point trajectories. In: ECCV, vol. V, pp. 282–295 (2010)

5. Dalal, N., Triggs, B.: Histograms of oriented gradients for human detection. In: Schmid, C., Soatto, S., Tomasi, C. (eds.) CVPR, pp. 886–893 (2005)

6. Davidson, P., Hautamaki, J., Collin, J.: Using low-cost mems 3d accelerometer and one gyro to assist gps based car navigation system. In: International Conference on Integrated Navigation Systems (2008)

7. Felzenszwalb, P., Huttenlocher, D.: Pictorial structures for object recognition. IJCV **61**(1), 55–79 (2005)

8. Felzenszwalb, P.F., McAllester, D.A., Ramanan, D.: A discriminatively trained, multiscale, deformable part model. In: CVPR (2008)

9. Fergus, R., Perona, P., Zisserman, A.: Object class recognition by unsupervised scale-invariant learning. In: CVPR, vol. II, pp. 264–271 (2003)

10. Ferrari, V., Jurie, F., Schmid, C.: Accurate object detection with deformable shape models learnt from images. In: CVPR (2007)

11. Gall, J., Lempitsky, V.: Class-specific hough forests for object detection. In: CVPR, pp. 1022–1029 (2009)

12. Huang, C., Wu, B., Nevatia, R.: Robust object tracking by hierarchical association of detection responses. In: ECCV, vol. II, pp. 788–801 (2008)

13. Kuhn, H.W.: The hungarian method for the assignment problem. Nav. Res. Logist. Q. 83–97 (1955)

14. Lampert, C., Blaschko, M., Hofmann, T.: Beyond sliding windows: object localization by efficient subwindow search. In: CVPR, pp. 1–8 (2008)

15. Lehmann, A., Leibe, B., Van Gool, L.: Fast prism: branch and bound hough transform for object class detection. IJCV **94**(2), 175–197 (2011)

16. Leibe, B., Schiele, B.: Interleaved object categorization and segmentation. In: BMVC, pp. 759–768 (2003)

17. Leibe, B., Schindler, K., Van Gool, L.: Coupled detection and trajectory estimation for multi-object tracking. In: ICCV, pp. 1–8 (2007)

18. Lowe, D.G.: Distinctive image features from scale-invariant keypoints. IJCV **60**, 91–110 (2004)

19. Maji, S., Berg, A.C., Malik, J.: Classification using intersection kernel support vector machines is efficient. In: CVPR (2008)

20. Maji, S., Malik, J.: Object detection using a max-margin hough transform. In: CVPR, pp. 1038–1045 (2009)

21. Mikolajczyk, K., Leibe, B., Schiele, B.: Multiple object class detection with a generative model. In: CVPR, vol. I, pp. 26–36 (2006)

22. Ohba, K., Ikeuchi, K.: Detectability, uniqueness, and reliability of eigen windows for stable verification of partially occluded objects. PAMI **19**(9), 1043–1047 (1997)

23. Okada, R.: Discriminative generalized hough transform for object dectection. In: ICCV, pp. 2000–2005 (2009)

24. Russakovsky, O., Ng, A.: A steiner tree approach to efficient object detection, pp. 1070–1077 (2010)

25. Schneiderman, H., Kanade, T.: Object detection using the statistics of parts. IJCV **56**(3), 151–177 (2004)

26. Stauffer, C., Grimson, W.E.L.: Adaptive background mixture models for real-time tracking. In: CVPR, pp. 2246–2252 (1999)

27. Viola, P.A., Jones, M.J.: Robust real-time face detection. IJCV **57**(2), 137–154 (2004)

28. Wang, Z., Kagesawa, M., Ono, S., Banno, A., Ikeuchi, K.: Global 3d modeling and its evaluation for large-scale highway tunnel using laser range sensor. In: ITS World Congress (2012)

29. Wojek, C., Roth, S., Schindler, K., Schiele, B.: Monocular 3d scene modeling and inference: understanding multi-object traffic scenes. In: ECCV, vol. IV, pp. 467–481 (2010)

30. Xue, L., Ono, S., Banno, A., Oishi, T., Ikeuchi, K.: Global 3d modeling and its evaluation for large-scale highway tunnel using laser range sensor. In: ITS World Congress (2012)

31. Yarlagadda, P., Monroy, A., Ommer, B.: Voting by grouping dependent parts. In: ECCV, vol. V, pp. 197–210 (2010)

32. Yeh, T., Lee, J., Darrell, T.: Fast concurrent object localization and recognition. In: CVPR, pp. 280–287 (2009)

Chapter 2

Vehicular Internet: Security & Privacy Challenges and Opportunities

Kamran Zaidi * and Muttukrishnan Rajarajan *

School of Mathematics, Computer Science and Engineering,
City University London, EC1V 0HB London, UK

ABSTRACT

The vehicular internet will drive the future of vehicular technology and intelligent transportation systems (ITS). Whether it is road safety, infotainment, or driver-less cars, the vehicular internet will lay the foundation for the future of road travel. Governments and companies are pursuing driver-less vehicles as they are considered to be more reliable than humans and, therefore, safer. The vehicles today are not just a means of transportation but are also equipped with a wide range of sensors that provide valuable data. If vehicles are enabled to share data that they collect with other vehicles or authorities for decision-making and safer driving, they thereby form a vehicular network. However, there is a lot at stake in vehicular networks if they are compromised. With the stakes so high, it is imperative that the vehicular networks are secured and made resilient to any attack or attempt that may have serious consequences. The vehicular internet can also be the target of a cyber attack, which can be devastating. In this paper, the opportunities that the vehicular internet offers are presented and then various security and privacy aspects are discussed and some solutions are presented.

KEYWORDS

vehicular internet; VANETs; networks; security; privacy; ad hoc networks

1. INTRODUCTION

The developments in the automotive industry in the last few decades have been impressive. Cars today are much more fuel-efficient than ever before. However, the advancements in automotive technology did not have the same impact on the safety of the roads. Vehicles today are still as vulnerable to accidents due to fog, ice, and other hazards on the road, but above all, they are vulnerable to human error. However, this is all set to change: the automotive industry has been working actively for years to put different sensors in cars and connect them to an on-board computer. With advancement in telecommunications, it is now possible to connect vehicles to each other through wireless technologies to enable them to communicate and cooperate. Now, not only is the automotive industry pursuing autonomous vehicles—they are being encouraged by the governments as well. The UK government announced in March 2015 that a £100 m funding for research into driver-less cars will bring in companies from not only the automotive industry but also from Information Technology (IT), telecommunications, and infrastructure [1]. In the US, car manufacturers like General Motors (GM) are already selling 4G Long Term Evolution (LTE)-connected cars in their 2015 fleet and they predict having fully autonomous cars by the end of this decade [2]. Moreover, IEEE believes that the need to get a driver's license might be eliminated by 2040 as autonomous cars would be ubiquitous [3].

Car manufacturers such as GM and Ford have opened up application development for their platforms by making their Application Program Interface (API) available to developers [4,5]. They plan to follow the conventional business model, *i.e.*, the developers submit their apps which are tested and approved before making them available for download. This LTE connectivity alone will serve to improve the in-vehicle infotainment services by providing access to high-speed internet, streaming movies, navigation, music, and live television, *etc.* The other aspect is commercial, *i.e.*, offering location-based services and ads to the vehicle passengers. The LTE connectivity will not only bring internet to the vehicle but also make the vehicle a part of the internet, easing the way for the Vehicular *Ad Hoc* Networks (VANETs).

The true potential of the connected vehicle will be realized only when vehicles are interconnected to each other. This network, formed by the interconnection of vehicles, is referred to as VANET. This paradigm shift in vehicular technology will usher in a new era of innovation and will open a huge range of application areas that can help in improving road safety and reducing accidents on roads.

VANETs are considered important due to their huge potential and numerous applications. VANETs not only offer immense safety enhancements but also many commercial opportunities. Wireless access in the vehicular environment (WAVE) is based on IEEE 802.11p standard and provides the basic radio standard for Dedicated Short Range Communication (DSRC) in VANETs. DSRC has been allocated 75 MHz in the 5.9 GHz band by the FCC (Federal Communications Commission) in the US, and a similar band has been allocated in Europe as well. Vehicles use DSRC radios to communicate with each other, *i.e.*, vehicle-to-vehicle (V2V) and vehicle-to-infrastructure (V2I) communication. The communication range of DSRC is between 300 and 1000 m.

With the addition of 4G/LTE connectivity in vehicles, it is obvious that connected vehicles would be able to share data through the cloud. The advantages of using cloud are huge as the opportunities for applications are virtually limitless. The communication environment with both LTE and DSRC is shown in Figure 1.

DSRC specifies one main type of beacon message, *i.e.*, Basic Safety Message (BSM), that a vehicle transmits every 100 milli secs. In BSM, vehicles broadcast their location, speed, acceleration, and other useful parameters, which help other vehicles avoid collisions and generate warnings for improved safety. These messages also help the vehicles maintain the list of their neighbours. The congestion information is already available through navigation systems (Sat Nav) but exact real-time position information of other vehicles is not. This is the future that can enable the driver to drive safely even if there is zero visibility or when the driver is unable to drive, *etc.* Therefore, the vision is to create a mutual awareness among all vehicles on the road that leads to driver-less cars.

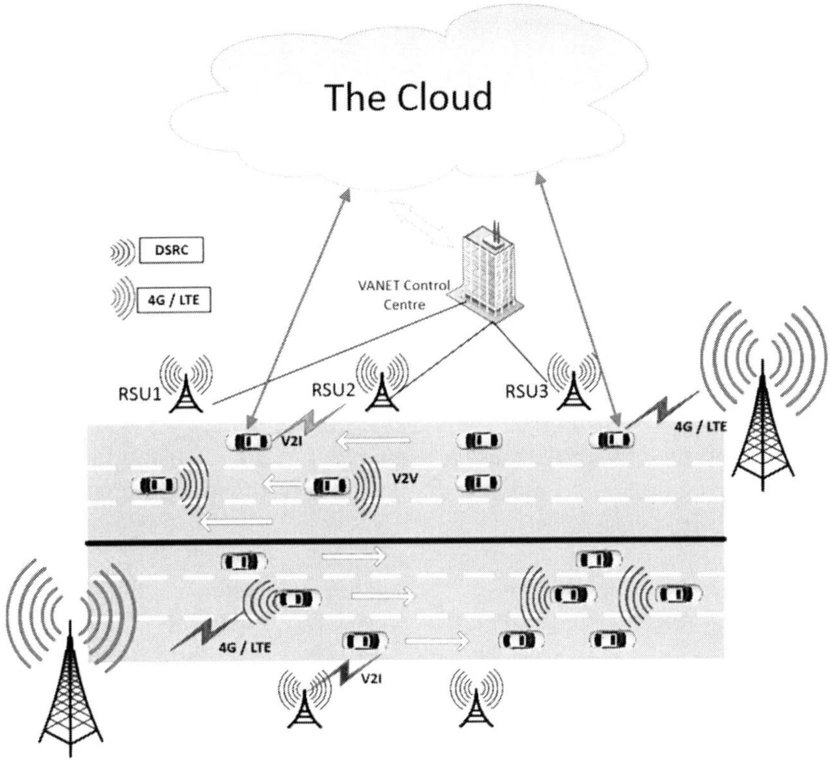

Figure 1. Vehicular Internet environment.

All vehicles can be programmed to transmit their location information whenever they make a sudden change in speed or position. However, the bandwidth in DSRC is quite limited, so we just cannot have vehicles transmitting their position and speed information all the time or we will have a jammed channel. The DSRC band consists of seven bands of 10 MHz each with a 5 MHz guard band in the beginning. One band out of these is reserved for BSM, but even this dedicated channel can be congested very quickly if the traffic is heavy and many vehicles are broadcasting BSMs.

The DSRC and 4G/LTE connectivity in vehicles can supplement each other and solve the bandwidth problem to give high quality, reliable, and secure information to VANET users. This combination will make it possible for a wide variety of applications and opportunities that will be discussed in the next section.

2. APPLICATIONS FOR VANETS

Many different convenience and commercial applications have been proposed for VANETs by researchers in [6,7,8] and safety applications have been proposed by the US Department of Transportation in [9]. With so many vehicles on the roads/highways having so much computing power and sensors, it is obvious and logical to utilize all this data to form an array of scattered sensor networks or highly mobile *ad hoc* networks. Some of the applications that have been recommended for VANETs include:

- **Safety applications:** Notifications for crashes, hazards on the roads (slippery or wet road conditions), traffic violation warnings, curve speed warnings, emergency electronics brake light, pre-crash sensing, co-operative forward collision warnings, *etc.* This could also include generating warning messages that inform drivers of approaching emergency vehicles.

- **Convenience applications:** Navigation, personal routing *etc.*, Congestion advice, toll collection, parking availability information, *etc.* Also, in disaster situations, the critical things are power failure and network breakdown. The connected vehicles can play a very significant role in such situations as they have on-board batteries and many sensors including cameras, *etc.*, thus providing valuable images and SOS calls. The vehicular network can become the emergency communication mechanism. Similarly, road and weather conditions can be monitored by sharing the data from on-board vehicle sensors [9].

- **Commercial applications:** Vehicle diagnostics exchanges for avoiding possible car problems, location-based services such as advertisements and entertainment, *i.e.*, data/video relay, social networking updates, *etc.*

In the applications discussed above, the vehicles are sharing information in V2V and V2I communications that includes their position information and also information of road conditions. Furthermore, the driver may need to make life-saving decisions based on the information received (e.g., emergency brake light, forward collision warning). It is therefore necessary that the privacy, reliability, and integrity of the messages are ensured so that actions can be taken in a split second after the information is received due to the speeds involved in the case of vehicles.

3. VANET AND THE CLOUD

The large amount of unused storage space and computing power coupled with on-board sensors provide a unique opportunity to utilize vehicles as a wireless sensor network. The vehicles will be equipped with not only sensors to measure temperature, humidity, wind, *etc.*, but also with cameras. This together with high speed (4G/LTE) connectivity can be a great advantage for a variety of applications. VANETs can be a part of the cloud in the following three possible ways (Figure 2).

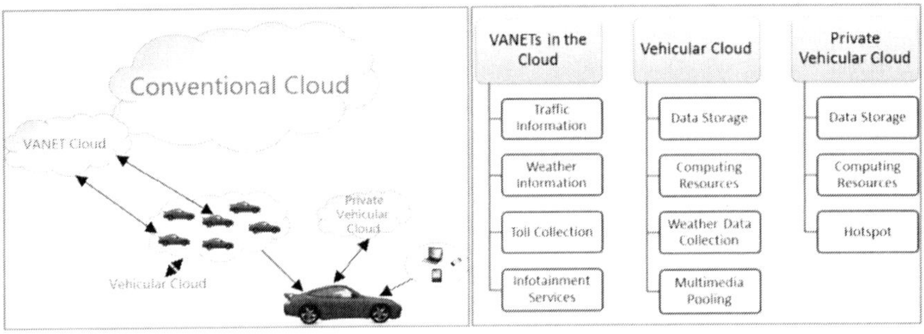

Figure 2. VANETs and cloud (left), cloud applications (right).

3.1. VANETs in the Cloud

The connected vehicles can connect to the cloud and upload their data, including sensor readings, traffic information, *etc.*, for downloading by other vehicles that are following behind. This allows the users to share data anonymously and preserve their security and privacy. The application servers in the cloud can aggregate the information coming in from vehicles and then compile the data to make announcements for traffic, road conditions, accidents, and weather conditions based on vehicle sensor data [9]. This improves the reliability of the information by filtering out biased or false information based on the number of users reporting such information.

3.2. Vehicular Cloud

The vehicles can connect together forming a vehicular cloud that will offer storage space, computing resources, sensor readings, *etc*. as an on-demand service to clients. The clients could be individuals, corporations, or government departments such as the meteorology department, police, or highway authorities. Moreover, with cameras onboard vehicles and high-speed connectivity, the police or emergency services can even request images from vehicles for an area that they want to monitor such as the site of an accident. Ericsson has already developed its own Connected Vehicle Cloud (CVC) for vehicle telematics and is behind Volvo Cars Sensus, which is an infotainment system for Volvo vehicles [10]. The CVC is designed to open new business opportunities by providing the vehicle data to companies, allowing developers to develop new applications by giving them an open platform, and facilitating the drivers by connecting them to the manufacturers and other service providers.

3.3. Private Vehicular Cloud

The storage, connectivity, and computing resources of the vehicle can be utilized by the owner as a private cloud, *i.e.*, the user connects to the internet and uploads their data to their "private cloud". This not only makes their data available to them whenever they want, but it also gives them physical ownership of their data.

4. AUTONOMOUS VEHICLE

The autonomous vehicle is a reality today (Google Car) and it is only logical to say that the technology will improve further and reduce in cost to become a feature in the near future. However, it will take some time for it to become a driver-less feature on a big scale. Also, it is possible that much like cruise control, this could become a feature that could be activated on the motorway/highway.

Autonomous vehicles would need almost no input from the driver. This would be possible with a lot of hardware like cameras, radars, and Laser Illuminated Detection and Ranging (Lidars) to sense oncoming obstacles. Image processing techniques are applied to images from the cameras to recognize and identify shapes and features. This enables the vehicle to sense the approaching obstacle and automatically apply breaks or reduce speed much more quickly than the driver would ever be able to. The use of these and other technologies therefore ensures that human errors can be reduced to a large extent and, as a result, roads can be made much safer. This would enable the driver to be more productive by doing something like reading, browsing on the internet or making a phone call. In terms of safety, autonomous vehicles are not only a possibility but a necessity for safe motoring. The next step in autonomous vehicles is vehicle swarm.

5. VEHICLE PLATOON/SWARM

Travelling at high speeds brings more problems into the picture and makes things a lot more complicated when the vehicle is on a motorway/highway. However, these problems can be reduced if the vehicles form a group and then travel in a formation on highways and motorways. This will have two benefits: first, vehicles are synchronized (*i.e.*, vehicles travel in a straight line and maintain a fixed distance from each other and the vehicle in front informs the cars behind if they should reduce speed or change lanes in advance, and then all of them can do it in tandem) and second, the randomness in the immediate vicinity of the vehicles is reduced, which makes things much simpler and safer. Safe Road Trains for the Environment (SARTRE) is an EU-financed 6.4-million-dollar project that has proven this concept in collaboration with Volvo and UK-based company Ricardo, among others [11]. The aim of the SARTRE project was to develop a road train or platoon concept in vehicular traffic where a lead vehicle driven by a professional driver takes control of a platoon of vehicles that go in a semi-autonomous state [11]. The SARTRE project demonstrated successfully that vehicle platoons are not only safe but can be very efficient in reducing congestion and improving safety.

The vehicle swarm (shown in Figure 3) can be imagined by thinking of the motorway/highway as a conveyor belt, but where a conveyor belt moves with a constant steady speed, a group of vehicles on the motorway can move with any set constant speed. This can be possible by allowing like-minded vehicles, *i.e.*, vehicles that intend to travel at roughly the same speed, to team up and form groups or a "vehicle swarm". Ricardo, a British engineering firm involved in the SARTRE project, believes that vehicle platooning is possible with fully autonomous vehicles [12].

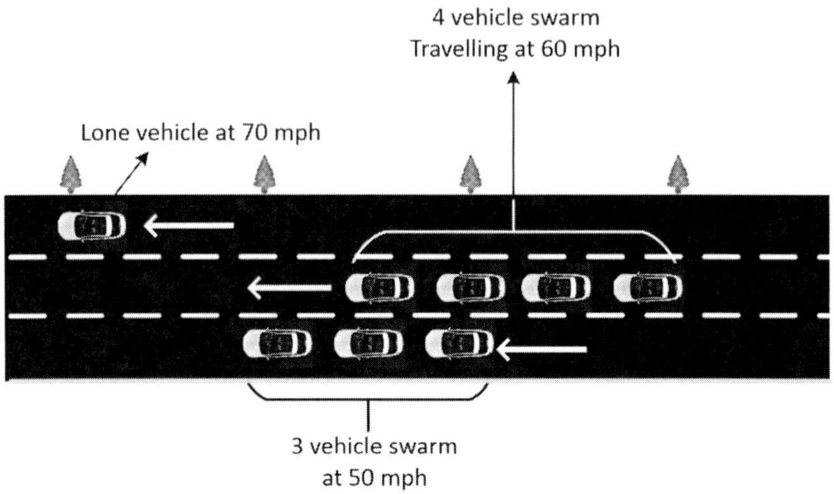

Figure 3. Vehicle swarm concept.

6. SECURITY AND PRIVACY ISSUES

Security and privacy of users in VANETs are very important but, at the same time, an objective and realistic definition of security and privacy is also needed. This means that the level of security and privacy assured in a VANET can be no more than it is in the real world [13]. In the real world, a vehicle can be identified by its registration plate and then tracked by authorities via video cameras as is the practice nowadays. Similarly, there are cases (e.g., stolen car, accident, or misbehaving vehicle) when the whereabouts of a vehicle in a VANET have to be determined by the authorities, e.g., in case of accidents, which implies traceability. The US Department of Transportation (DOT) has identified many safety applications which require each vehicle to periodically transmit its current location to its one-hop neighbours at a rate of up to 10 Hz [14]. These messages, called Cooperative Awareness Messages (CAMs), are susceptible to an eavesdropper who can then get the location of that vehicle. Therefore, it is necessary that the messages are secured from giving away the user's location. The security and privacy goals and how they are achieved in VANETs are shown in Figure 4 and are discussed below in detail.

It is essential for VANET security that the location of a vehicle is transmitted while preserving the location privacy of the user. This location information can be used by adversaries in tracking vehicles or routes. As discussed previously, location privacy requires unlinkability, which ensures anonymity. As the information in VANETs is very critical for safety, it is imperative that it should be reliable. Reliability is achieved through authentication, which ensures the integrity of the messages, *i.e.*, they have not been tampered with. Another requirement for VANETs is non-repudiation, which means that users should not be able to deny sending a message so that they can be tracked and penalized in case of a false message. This is achieved by making the messages traceable but only by the authorities so that they can be revoked. It has been proposed [15] that the tracing should be done by multiple authorities in order to provide extra security and privacy. Different security schemes have been used to protect privacy or achieve anonymity and, based on the underlying security mechanisms they use, these schemes can be classified as:

- Pseudonyms coupled with public key infrastructure (PKI)
- Trust-based schemes
- Group signatures
- k-anonymity schemes
- Identity-based signature schemes

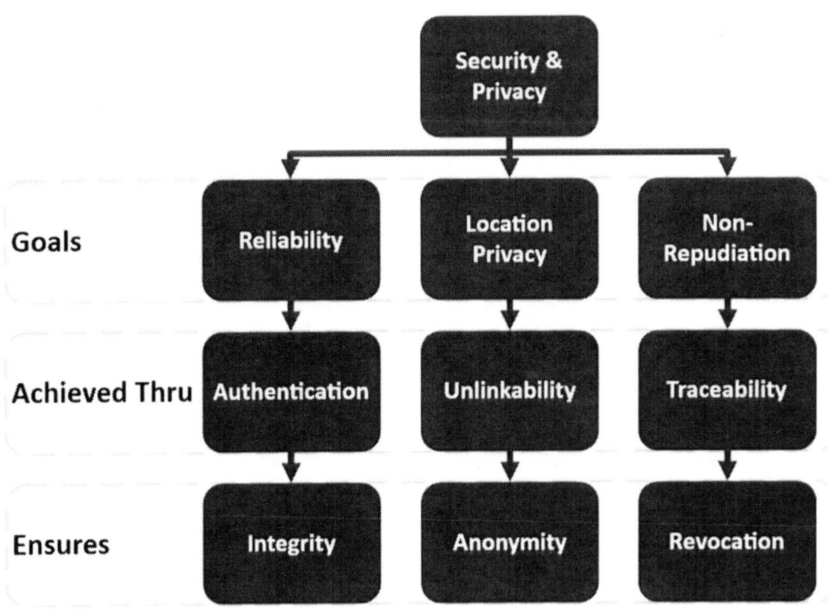

Figure 4. Security and privacy goals in VANETs and how they are achieved.

There are various issues associated with each of these schemes which are discussed in the next section.

7. REVIEW OF SECURITY AND PRIVACY-PRESERVING SCHEMES IN VANETS

7.1. Pseudonyms Coupled with PKI-Based Schemes

As the vehicles are transmitting their location information, e.g., to warn other users of accidents, researchers [16,17] have discussed using pseudonyms to preserve the privacy of users. The use of Pseudonyms (PNs) hides the real identity of the user and prevents the linking of the real identity of the user with the pseudonym, thereby providing unlinkability. However, the Regional Trusted Authority (RTA) that issues pseudonyms can reveal the real identity of the user if the user commits an illegal act, hence providing traceability.

In all cases, the On Board Unit (OBU) is assumed to be a tamper-proof device which contains the unique Vehicle Identification Number or VIN. This VIN is tied to an identity certificate at the time of registration. From here on, the Trusted Authority (TA) (e.g., transport authority) is tasked with issuing new blocks of pseudonyms/certificates to the vehicles, to be used while communicating with the Road Side Units (RSUs). These certificates have a validity period and the vehicle switches its pseudonyms to ensure privacy. The authors in [18] suggest that short-term linkability should be allowed so that the receiver is able to verify that two or more messages in a short time frame have come from the same sender, as this might be required by applications. The recommended time after which the pseudonyms are switched also varies, e.g., every minute or every few minutes, with every message or every second [19]. However, the frequency of PN changes is directly proportional to the computation overhead at the RSU, which can slow down packet delivery. Also, the OBU is also supposed to store the public keys of all CA (Certificate Authority) / TA. Furthermore, different PKI-based techniques that use a CA/TA connected with the RSU have been suggested.

The downside of the pseudonym scheme is that the user (OBU) has to have hundreds or thousands of these pseudonyms stored on board and they need to be refreshed after some time (e.g., once a year). Different papers have explored this and proposed different methods of refreshing these, e.g., downloading new pseudonyms when the vehicles go in for service (once a year). It is also suggested that the pseudonyms are refilled at social spots [20]. In [13], a PN distribution scheme is proposed which incorporates control channels and service channel intervals in communication between vehicle and RSU for PN refill. It is important to note that pseudonym schemes alone do not support authentication, integrity, and non-repudiation. Signing protocols have been proposed that provide integrity and authentication. In this, a large number of certified public-private key pairs are stored in the OBU. Each key pair will be used for a short period of time and then discarded. Using PKI involves downloading and maintaining public key certificates, which results in heavy computational overhead that is not desirable in VANETs as the computing power of OBU is limited.

Anonymous certificates in PKI-based schemes can guarantee identity privacy but cannot guarantee location privacy. This is because an attacker can monitor the certificate change by a vehicle between two observation points while moving in the same lane with the same speed. Also, the TA/CA has the ability/capability to identify the real identity of a vehicle based on its anonymous certificate. In [15], a method has been proposed so that multiple authorities are required to de-anonymize a user to increase the security. In [17], authors propose that the users change their PNs only at predetermined locations (mix zones) where the density of vehicles is high and the speed and direction of vehicles changes often. However, the authors conclude that such a technique provides limited privacy due to the inherent lack of randomness in vehicle mobility.

A major drawback of using PN is maintaining the Certificate Revocation List (CRL), which keeps track of revoked certificates of misbehaving vehicles. This list then has to be checked to ensure whether a vehicle is revoked or not, which is both time- and resource-consuming. Authors in [19] propose a reduction in CRL size by limiting the list to regions, *i.e.*, each RTA will maintain its own CRL, thereby reducing the size of each CRL.

7.2. Trust-Based Schemes

VANETs share many characteristics with Mobile *Ad Hoc* Networks (MANETs). Trust-based routing schemes have been proposed for MANETs and the same idea of trust has also been applied to VANETs in [16,21,22]. However, apart from a lot of similarities between them, e.g., decentralized system, mobility, and openness, VANETs are different as they consist of a much larger number of nodes and their topology changes quickly as vehicles move fast [23]. Therefore, forming a network based on the trustworthiness of vehicles is quite challenging. In MANETs, the focus is on reliable packet delivery, whereas in VANETs the aim is to increase road safety and, therefore, the decision-making process has to be very fast due to the high speeds and limited time involved.

In [24], authors propose trust establishment through infrastructure or in a self-organizing manner. The former means to use a central authority or security infrastructure and the latter involves developing a trust score dynamically. In infrastructure-based trust establishment, there has to be a centralized authority, but it is very difficult to do that quickly in VANETs due to the size and time restrictions. In self-organizing trust establishment, the trust is based on the direct (self-interaction), indirect (receiving interaction information from other nodes), and hybrid approaches (combination of the two) as shown in Figure 5.

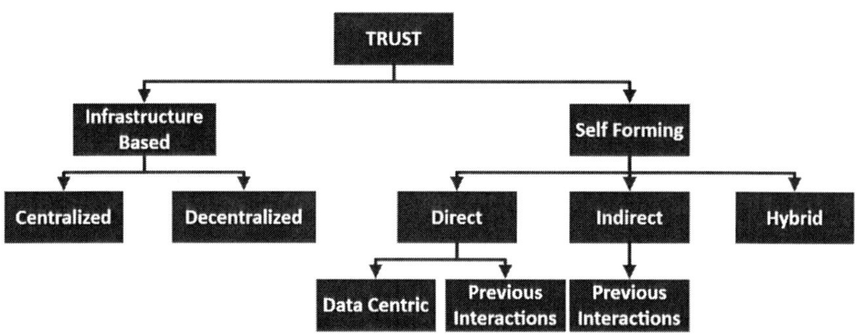

Figure 5. Classification of trust establishment approaches.

As the vehicle's trust score has to be updated and linked with some identification (Pseudonym/Certificate, *etc.*), it again raises the problems of privacy. Moreover, in VANETs, the trust information might not be available in the beginning, or a default trust score has to be assigned. Also, once a group of vehicles form a network, it is then difficult to protect oneself from an insider. To account for this, data-centric trust schemes have been suggested in [25], where a malicious node can be identified based on the data/information being exchanged. In [26], a data-centric mechanism is proposed that enables users to aggregate the data and then determine if a rogue is sending false data. The rogue node can then be reported and its data is ignored.

7.3. Group Signatures

For VANETs, authors in [27] proposed the grouping of vehicles travelling at the same speed and in the same direction. This allows group members to anonymously issue messages and sign them with the group signature on behalf of the group. This also allows extended silent periods for vehicles as only one vehicle in the network transmits at a time. By combining the vehicles into groups, the authors proposed that a vehicle can reduce its V2I transmission, which enhances the anonymity of vehicles. Group signatures allow the distribution of a group public key associated with multiple group private keys. Therefore, an attacker will be able to detect that a message has been sent by a particular group but it will not be able to identify which vehicle sent the message [24].

In [18], a scheme based on group signatures and short-lived keys, called Temporary Anonymous Certified Keys (TACS), is presented as an efficient way to fulfill the privacy and safety requirements. The scheme suggests a long-term group key stored in the OBU provided by a managing authority which the OBU uses anonymously to prove to the RA that it is a valid OBU and to get a short-term certified key that is only valid in that RA's region. As soon as the OBU moves into a new region, it has to update its TACS by getting a new short-lived certified key from the new RA. This ensures that OBUs update their keys after entering a new region. This ensures short-term linkability and traceability in the long run as the managing authority can be queried by the RA for the real identity of the vehicle.

Group signatures provide a high level of privacy but revocation becomes a serious problem when the size of the group increases, thereby raising scalability issues. Furthermore, they can be computationally more expensive [28].

7.4. Identity-Based Encryption Schemes with Pseudonyms

Identity-based encryption (IBE) was first proposed in [29]. Identity-based cryptography allows the public key of an entity to be derived from its public identity information such as name, email address, or VIN, *etc*. Anybody who wants to send a message to a user can use the recipient's identity to get their public key and encrypt the message using this key, which can only be decrypted by the private key of the intended recipient. A private key generator (PKG) is used to generate and distribute private keys for users (through a secure channel) and also to distribute the public key. IBE is highly suited to a dynamic *ad hoc* network like VANETs, as it allows nodes that have not met before to start communicating with each other quickly and safely.

The authors in [15,30,31] have used the idea of IBE in VANETs. The proposed architecture in [31] consists of four entities, two TAs which are trace authority (TRA), a PKG, an RSU at the roadside, and mobile OBUs on vehicles. Privacy in IBE schemes is achieved by using pseudonyms that the user can request from the TRA. The TRA is responsible for registration of RSUs and OBUs, and can reveal the actual identity of the signing OBU. The PKG is responsible for generating and assigning private keys for OBUs and RSUs. Self-generating pseudonyms (PNs) are suggested for privacy preservation. For authentication, a pool of PNs is preloaded into a vehicle for different regional trusted authorities (RTAs). Users from different regions can authenticate each other via RTAs. RTAs are responsible for generating cryptographic key materials for the RSUs and the vehicles in the region and deliver them over secure channels. The users use their self-generated pseudonyms as identifiers instead of real-world identities. Similarly, the idea of IBE is used in [30] to improve the performance and reduce the processing time of RSU when it is verifying signatures for a large number of users.

Identity-based cryptography seems to be a good candidate for security and privacy in VANETs but has its own limitation and challenges, especially with revocation, which is still a major problem and is open for research. Secure channels for secret key distribution and heavy computation costs are other factors to consider.

7.5. k-Anonymity Schemes

Researchers have proposed k-anonymity schemes for VANETs. In [32], a k-anonymity scheme has been proposed where k vehicles in a region are assigned the same PN for communicating with the RSU. This ensures that an attacker cannot associate a message with a specific vehicle. An attacker can only detect that a group of nodes are receiving the messages but cannot determine which one in particular. The source and destination anonymity cannot be guaranteed as VANETs are inherently changing all the time, therefore, it is difficult to identify the source and destination. In k-anonymity schemes, when a message has to be delivered to a vehicle, the region is flooded with the message to ensure that it reaches the destination vehicle. However, in such a multi-hop scheme there is a clear problem of the scarce bandwidth resource utilization. The network might be flooded by the same message, causing congestion which is highly undesirable in VANETs.

8. PERFORMANCE COMPARISON

In order to compare the performance of different schemes discussed in this paper, it is necessary to have complete details of the algorithms proposed for all the schemes. One parameter that can be used to judge the performance of networks is control overhead, but in order to calculate this, it is necessary to know the exact type and size of data that the OBU will be handling. This can only be obtained after finalizing the type of applications that will be deployed in VANETs. Similarly, bandwidth utilization is another important parameter that has to be taken into account, but it is difficult to calculate it as the type of data and applications in VANETs have not been finalized. Therefore, we have compared the major performance characteristics of the discussed schemes by defining or modifying some parameters in Table 1 and we have discussed the major features and disadvantages of each scheme in Table 2. For comparison of the schemes discussed, we have used the rating system of HIGH, MED, and LOW in Table 1. The performance metrics used in Table 1 are described below:

- **Scalability:** Scalability is a well-defined performance metric for network protocols and architectures and it means how well a system can cope with the expansion of the network while maintaining the performance standards. A LOW in this category means that the scheme is not suitable for a network in which the number of nodes can grow beyond a small number. A MED in this category means that the scheme can work well for a limited number of nodes but it should not exceed that limit. A HIGH means that the scheme is highly scalable and can work well for a very large number of nodes.

- **Computing cost:** The computing cost has been chosen as a metric because it is always in demand and usually it is always less than what is desired. Furthermore, experience has shown that the appetite of programs and data eats up the computing power available very quickly. We have considered the computing cost at both the RSU and OBU. This metric is especially important at the OBU end where we can have reasonable, but never excessive, computing power. A LOW here means that the computing resources required are much less than available. HIGH means that computing resources available might fall short of what is required. MED means that the computation resources are sufficient.

- **Privacy:** This is the fundamental requirement of all schemes studied and has been rated according to the merits and demerits of each when viewed in its entirety, e.g., in the case of pseudonym *cum* PKI schemes, the privacy has been rated as MED as the PNs changing and their maintenance poses a serious challenge for the successful deployment in VANETs. HIGH means strong, MED means acceptable but risky, and LOW means unacceptable or no level of privacy.

- **Latency:** This is another well-known parameter for networks that means delays which are experienced in a network due to any reason. A network with LOW latency is considered to be fast and vice versa.

- **Cost of deployment:** This parameter indicates whether the infrastructure requirements make it easy to deploy in the practical world or not. A scheme with HIGH cost of deployment is obviously high-cost and difficult to deploy, *i.e.*, with many additions/changes to the existing network making it impractical. A scheme which gets a LOW in this metric will be low-cost to deploy and will require a few changes to the existing network to make it feasible.

Table 1. Performance metrics for security schemes in VANETs.

S/No	Scheme	Scalability	Computing cost		Privacy	Latency	Deployment cost
			RSU	OBU			
1	Pseudonym *cum* PKI	HIGH	HIGH	HIGH	MED	MED	HIGH
2	Group Signatures	LOW	MED	HIGH	HIGH	HIGH	HIGH
3	Trust-Based Schemes	LOW	HIGH	HIGH	LOW	HIGH	HIGH
4	K-anonymity	LOW	MED	HIGH	HIGH	HIGH	MED
5	Identity-Based Encryption (IBE) with Pseudonyms	HIGH	MED	HIGH	HIGH	HIGH	HIGH

Table 2. Qualitative Comparison of Security Schemes in VANETs.

S/No	Schemes	Features	Disadvantages
1	Pseudonym *cum* PKI	(*a*) User privacy is achieved by using pseudonym coupled with PKI; (*b*) Certificates (public, private key pairs) are downloaded from trusted authority; (*c*) Pseudonyms are changed continuously for preserving privacy.	(*a*) Thousands of certificates to be downloaded to OBU; (*b*) Certificates need to be replenished periodically; (*c*) CRL has to be maintained and it keeps on changing and is time- and resource-hungry; (*d*) The CA can link pseudonyms with vehicles, and therefore have to be secure.
2	Group Signatures	(*a*) Privacy achieved by forming groups; (*b*) Reduced transmission by ways of periodic broadcasts by a single member of group.	(*a*) TA can reveal the real identity of user (*b*) CRL has to be maintained and checked and it increases with the size of the group.
3	Trust-Based Schemes	Information accepted based on trust. Trust established based on previous record in a centralized authority or based on current and previous interactions with the user in the same session.	(*a*) No initial trust information available as centralized system would be too slow/too huge; (*b*) Protection against inside attackers is difficult; (*c*) Trust score has to be maintained and checked.
4	K-anonymity	(*a*) Messages are disseminated by ways of flooding it to neighbours; (*b*) Ensures privacy as long as size of group is adequate.	Flooding is used to disseminate messages which if not effective, can overwhelm the network and eat up bandwidth for the same message. Efficient and practical algorithm missing.

5	Identity-Based Encryption (IBE) With Pseudonyms	(*a*) User's public key is derived from his public identification such as VIN, *etc*., which eliminates the need for public key distribution (*b*) No need for certificate downloads and storage in OBUs (*c*) Additional information such as a time stamp can also be added.	(*a*) Secure channel needed for private key distribution; (*b*) PKG has to be highly secure; (*c*) TRA can reveal the real identity of the user, therefore must be secure; (*d*) Revocation is still an open problem.

9. CHALLENGES

9.1. Security

Securing the vehicle from external attacks so that vehicle is not hijacked remotely is a very real threat and will be the most important challenge to address. This will be especially challenging in autonomous or driver-less vehicles, as by design the vehicle is required to navigate on its own by controlling the steering and control system of the vehicle. The brains of the vehicle or the OBU will control the movement of the vehicle and the OBU will be connected to the internet and the vehicular cloud. Therefore, securing the control systems of the vehicle will be the first thing that has to be done and is only possible with very strong security systems in place to secure the vehicles from a cyber attack.

9.2. Privacy

The information that is collected from the car's sensors has to be protected and should only be used after anonymizing it. The concern is that your car should warn you if you are over the speed limit but should not turn against you, *i.e.*, reporting to the authorities that you went over the speed limit. However, there are many instances when sharing location information with some service providers might actually be beneficial for the user, e.g., insurance companies might offer to reduce the insurance premiums if such details are shared. There can be similar incentives from other service providers to use vehicle sensor data in return for some reward that can actually encourage participation from the users.

9.3. Cost

Another important factor is undoubtedly the cost. When there are so many features and hardware available for the vehicles, it all comes down to how affordable all this can be made. A logical solution could be offering all the available features to the buyer as a customizable solution at the time of buying the vehicle or as later add-ons, which will keep the cost of the vehicle in check.

10. POSSIBLE SOLUTIONS TO SECURITY ISSUES

We have proposed some solutions in [15,26] to the issues mentioned above. The issues of identity, PN generation, security, and privacy preservation have been addressed in [15]. We propose a digital identity scheme for VANETs (DIVA) by considering that it is the driver who is to be held liable for the vehicle and not *vice versa*. Therefore, we use the identification of the vehicle (VID) and the driver (DrL) to form a new digital identity. The two identities are merged into one after authentication from two authorities, *i.e.*, the vehicle registration authority and the driver license authority. Each of the two authorities issues the user a token and secret keys which are then used by the user to generate their pseudonym. Therefore, even if the master key is compromised, the tokens can only be traced by the two authorities and no individual authority can reveal the true identity of the user. Moreover, in DIVA, the PNs are being generated at the OBU, which reduces the latency at the RSU. The performance of the scheme is compared with two other schemes efficient conditional privacy preservation (ECPP) [33] and pseudonymous authentication-based conditional privacy (PACP) [34] as shown in Figure 6.

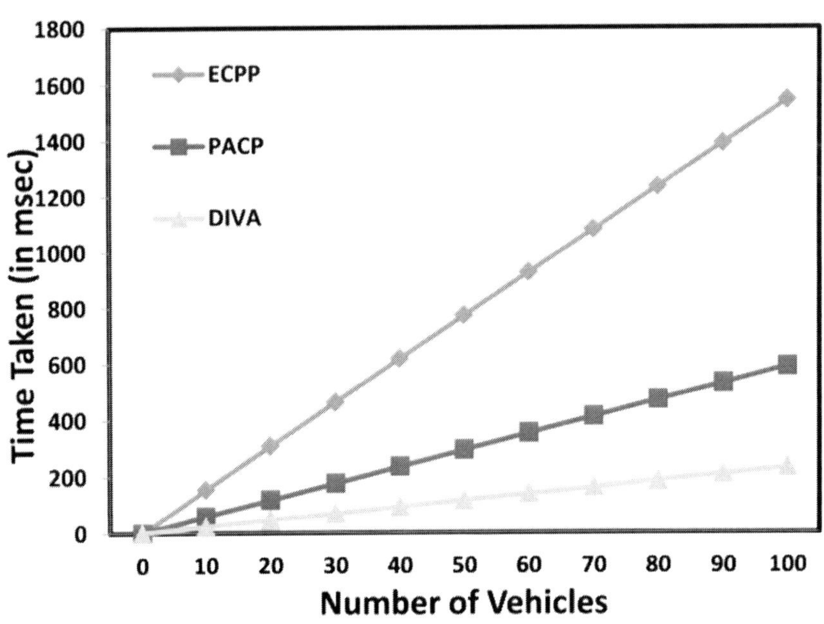

Figure 6. Comparison of protocol latency at RSU [15].

To compute latency at the RSU, the same parameters are used as in [34], *i.e.*, key size of 160 bits and times required for a pairing operation and point multiplication to be 7.3 ms and 8.5 ms, respectively. The major operations in our scheme for RSU to perform decryption and signature verification involve two pairing operations (PO) and one point multiplication (MP), whereas ECPP requires 13 MPs and six POs and PACP requires two POs and one MP for verification at RSU. However, in both PACP and ECPP, the RSU generates and verifies the PN tokens and short-lived key pairs, respectively, for the vehicles, which causes more delay as shown in Figure 6. The graph shows the latency when only one token is being generated for each vehicle as the number of vehicles increases from one to 100. In DIVA, the PNs are being generated at the OBU and therefore, there is no extra delay at the RSU and the latency is reduced at the RSU, which is desirable.

The issues associated with trust and rogue nodes in VANETs are addressed in [26]. As discussed previously, trust is difficult to maintain in VANETs and honest nodes can turn rogue once they become part of the network. Therefore, a data-centric rogue node detection scheme (C-DAC) is presented in [26] to address the issues of trust and security in VANETs. A mechanism has been proposed where users can aggregate the data that they are receiving while sharing their own data with their neighbours. The scheme is not dependent on any past trust value and only relies on the data that is received from the node, which is highly desirable as trust is difficult to manage in VANETs. C-DAC allows the vehicles to get a very good picture of the *normal* conditions/data in their vicinity. Therefore, any abnormal or anomalous data received can be rejected and the user reported. In order to check the proposed model, it is simulated using OMNET++, SUMO [35], and VACaMobil [36]. OMNET is a modular C++ library and framework that is used for network simulations. Simulation of Urban Mobility (SUMO) is a software tool for vehicular traffic generation by specifying speed, vehicle behaviour, road type, and conditions. VACaMobil is a car mobility manager for OMNET that works with SUMO. Different scenarios with and without accidents are simulated with parameters given in Table 3. The results show that using this scheme, vehicles can effectively determine false data and the emergency message can successfully be conveyed to vehicles that are far off so that they can gradually slow down.

The proposed C-DAC scheme also effectively solves the problem of broadcast storms, *i.e.*, when nodes try to send the same emergency message to alert other vehicles causing congestion and a choked network very quickly. C-DAC has the ability to convey the emergency information to other vehicles far behind successfully and efficiently without causing network congestion. This success, *i.e.*, how far the information is successfully received on the road, is compared with another scheme, Adaptive and Mobility Based Algorithm (AMBA) [37], and is shown in Figure 7.

Table 3. Simulation parameters for C-DAC [26].

Parameter	Value
Simulation Time	500 s
Scenario	3 Lane Highway
Highway Length	5-Kms
Max Vehicle Speed	28 m/s or 100 Km/h
Mobility Tool	VACaMobil
Network Simulation Package	OMNET++
Vehicular Traffic Generation Tool	SUMO
Number of Vehicles	330
Vehicle Density	20–30 veh/Km
Wireless Protocol	802.11p
Transmission Range	500 m in each direction

11. CONCLUSIONS

It is evident from the discussion above that the security and privacy challenges in vehicular internet are significant. The risks associated with the adoption of technologies associated with VANETs will be weighed against the benefits they provides. However, the benefits clearly outweigh the risks and this is what the users will consider when adopting these technologies. The privacy will be important to the user but probably not in every case and not for everyone. Also, the incentives will decide whether the users voluntarily share the data or share it after anonymizing it. The autonomous car is a reality today and there is no doubt that it will become commonplace in the years to come. The connected vehicle is imminent and it is important that steps are taken to minimize the security and privacy issues associated with it before its launch.

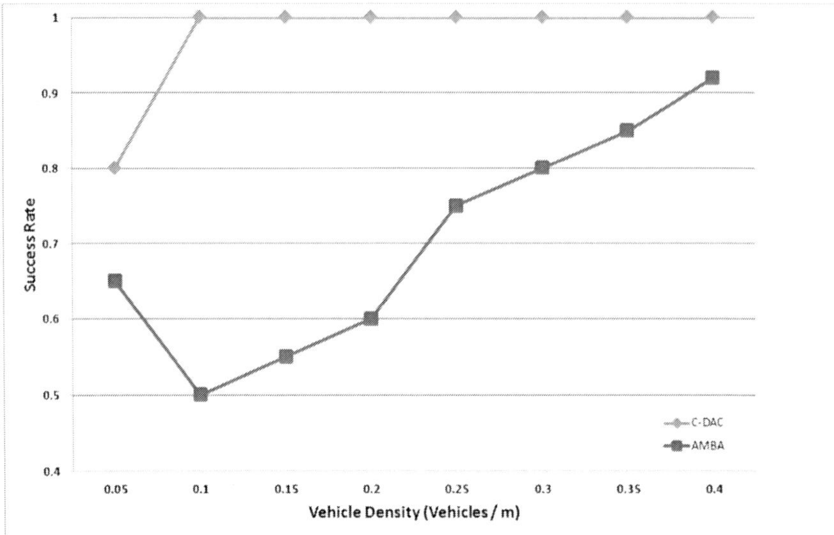

Figure 7. Percentage of vehicles within 3 km that received the emergency information successfully [26].

AUTHOR CONTRIBUTIONS

Kamran Zaidi contributed to the design of the proposed schemes, implementation and simulation setup and writing of the initial draft paper. Muttukrishnan Rajarajan supervised the research work and provided guidance in terms of the structure, content and proofreading of the final paper.

REFERENCES

1. The Telegraph. Available online: http://tinyurl. com/ telegraph- co-uk-finance-budget (accessed on 10 June 2015).

2. General Motors. Available online: http://media. gm.com/media/us/en/ gm/ news. detail.html/content/ Pages/news/us/en/2013/Feb/ 0225_ 4g-lte.html (accessed on 10 June 2015).

3. IEEE Intelligent Transportation System Society. Available online: http://its.ieee.org/2014/09/15/you-wont-need-a-drivers-license-by-2040/ (accessed on 07 June 2015).

4. Ford. Available online: https://developer.ford.com/ (accessed on 09 June 2015).

5. General Motors. Available online: https://developer.gm.com/ (accessed on 10 June 2015).

6. Hartenstein, H.; Laberteaux, K.P. A Tutorial Survey on Vehicular Adhoc Networks. *IEEE Commun. Mag.* **2008**, *46*, 164–171.

7. Faezipour, M.; Nourani, M.; Saeed, A.; Addepalli, S. Progress and Challenges in Intelligent Vehicle Area Networks.*Commun. ACM* **2012**, *55*, 90–100.

8. Haas, J.J.; Hu, Y.; Laberteaux, K.P. Real-World VANET Security Protocol Performance. In Proceedings of the IEEE GLOBECOM 2009, Global Telecommunications Conference, Honolulu, HI, USA, 30 November–4 December 2009.

9. United States Department of Transportation. Available online: http://www.its. dot.gov/connected_ vehicle/connected_ vehicle_ apps. htm (accessed on 09 June 2015).

10. Ericsson. Available online: http://archive.ericsson.net/service/internet/picov/get?DocNo=28701-FGD101192&Lang=EN&HighestFree=Y (accessed on 08 June 2015).

11. The SARTRE Project. Available online: http://www.sartre-project. eu/en/Sidor/default.aspx (accessed on 09 June 2015).

12. New Scientist. Available online: http://tinyurl.com/newscientist-com-article (09 June 2015).

13. Benin, J.; Nowatkowski, M.; Owen, H. Unified pseudonym distribution in VANETs. In Proceedings of the 2010 IEEE 6th International Conference on Wireless and Mobile Computing, Networking and Communications (WiMob), Niagara Falls, ON, USA, 11–13 October 2010; pp. 529–533.

14. *Vehicle Safety Communications Project Final Report: Identify Intelligent Vehicle Safety Applications Enabled by DSRC*; DOT HS 809 859; U.S.

Department of Transportation, National Highway Traffic Safety Administration: Washington, DC, USA, 2005.

15. Zaidi, K.; Rahulamathavan, Y.; Rajarajan, M. DIVA-Digital Identity in VANETs: A multi-authority framework for VANETs. In Proceedings of the 2013 19th IEEE International Conference on Networks (ICON), Singapore, Singapore, 11–13 December 2013; pp. 1–6.

16. Abumansoor, O.; Boukerche, A. Towards a Secure Trust Model for Vehicular Ad Hoc Networks Services. In Proceedings of the 2011 IEEE Global Telecommunications Conference (GLOBECOM 2011), Houston, TX, USA, 5–9 December 2011; pp. 1–5.

17. Liu, B.; Chiang, J.; Haas, J.; Hu, Y. Short paper: A practical view of "mixing" identities in vehicular networks. In WiSec'11, Proceedings of the Fourth ACM Conference on Wireless Network Security; ACM: New York, NY, USA, 2011; pp. 157–162.

18. Studer, A.; Shi, E.; Bai, F.; Perrig, A. TACKing Together Efficient Authentication, Revocation, and Privacy in VANETs. In Proceedings of the 6th Annual IEEE Communications Society Conference on Sensor, Mesh and Ad Hoc Communications and Networks (SECON'09), Rome, Italy, 22–26 June 2009; pp. 1–9.

19. Benin, J.; Nowatkowski, M.; Owen, H. Framework to Support Per Second Shifts of Pseudonyms in Regional VANETs. In Proceedings of the 2010 IEEE 72nd Vehicular Technology Conference, Fall (VTC 2010-Fall), Ottawa, ON, USA, 6–9 September 2010; pp. 1–5.

20. Lu, R.; Lin, X.; Luan, T.H.; Liang, X.; Shen, X. Pseudonym Changing at Social Spots: An Effective Strategy for Location Privacy in VANETs. *IEEE Trans. Veh. Technol.* **2012**, *61*, 86–96.

21. Haas, J.J.; Hu, Y.-C.; Laberteaux, K.P. Design and analysis of a lightweight certificate revocation mechanism for VANET. In Proceedings of VANET; ACM: New York, NY, USA, 2009; pp. 89–98.

22. Raya, M.; Papadimitratos, P.; Aad, I.; Jungels, D.; Hubaux, J.-P. Eviction of misbehaving and faulty nodes in vehicular networks. *IEEE J. Sel. Areas Commun.* **2007**, *25*, 1557–1568.

23. Zhang, J. A Survey on Trust Management for VANETs. In Proceedings of the 2011 IEEE International Conference on Advanced Information

Networking and Applications (AINA), Biopolis, Singapore, 22–25 March 2011; pp. 105–112.

24. Wex, P.; Breuer, J.; Held, A.; Leinmuller, T.; Delgrossi, L. Trust Issues for Vehicular Ad Hoc Networks. In Proceedings of the Vehicular Technology Conference, 2008, Singapore, Singapore, 11–14 May 2008; pp. 2800–2804.

25. Kargl, F.; Papadimitratos, P.; Buttyan, L.; Muter, M.; Schoch, E.; Wiedersheim, B.; Thong, T.-V.; Calandriello, G.; Held, A.; Kung, A.; *et al.* Secure vehicular communication systems: Implementation, performance, and research challenges.*IEEE Commun. Mag.* **2008**, *46*, 110–118.

26. Zaidi, K.; Milojevic, M.; Rakocevic, V.; Rajarajan, M. Data-centric Rogue Node Detection in VANETs. In Proceedings of the 2014 IEEE 13th International Conference on Trust, Security and Privacy in Computing and Communications (TrustCom), Beijing, China, 24–26 September 2014; pp. 398–405.

27. Sampigethaya, K.; Huang, L.; Li, M.; Poovendran, R.; Matsuura, K.; Sezaki, K. CARAVAN: Providing location privacy for vanet. In Proceedings of the Workshop on Embedded Security in Cars (ESCAR), Cologne, Germany, 29–30 November 2005.

28. Ma, D.; Tsudik, G. Security and privacy in emerging wireless networks [Invited Paper]. *IEEE Wirel. Commun.* **2010**, *17*, 12–21.

29. Shamir, A. Identity-based cryptosystems and signature schemes. In *Advances in Crypto'84*; Lecture Notes in Computer Science; Springer-Verlag: Berlin, Germany; Heidelberg, Germany, 1984; Volume 196, pp. 47–53.

30. Shim, K. CPAS: An Efficient Conditional Privacy-Preserving Authentication Scheme for Vehicular Sensor Networks.*IEEE Trans. Veh. Technol.* **2012**, *61*, 1874–1883.

31. Lu, H.; Li, J.; Guizani, M. A Novel ID-based Authentication Framework with Adaptive Privacy Preservation for VANETs. In Proceedings of the Computing, Communications and Applications Conference (ComComAp), Hong Kong, China, 11–13 January 2012.

32. Zhang, C.; Lin, X.; Lu, R.; Ho, P. RAISE: An efficient RSU-aided message authentication scheme in vehicular communication networks. In Proceedings of the IEEE International Conference on Communications (ICC'08), Beijing, China, 19–23 May 2008; pp. 1451–1457.

33. Lu, R.; Lin, X.; Zhu, H.; Ho, P.H.; Shen, X. ECPP: Efficient Conditional Privacy Preservation Protocol for Secure Vehicular Communications. In Proceedings of the IEEE 27th Conference on Computer Communications, Phoenix, AZ, USA, 13–18 April 2008; pp. 1229–1237.

34. Dijiang, H.; Misra, S.; Verma, M.; Xue, G. PACP: An Efficient Pseudonymous Authentication-Based Conditional Privacy Protocol for VANETs. *IEEE Trans. Intell. Transp. Syst.* **2011**, *12*, 736–746.

35. Behrisch, M.; Bieker, L.; Erdmann, J.; Krajzewicz, D. Sumo-Simulation of Urban Mobility: An overview. In Proceedings of the SIMUL 2011, the 3rd International Conference on Advances in System Simulation, Barcelona, Spain, 23–29 October 2011; pp. 63–68.

36. Baguena, M.; Tornell, S.; Torres, A.; Calafate, C.; Cano, J.-C.; Manzoni, P. Vacamobil: Vanet car mobility manager for omnet++. In Proceedings of the 2013 IEEE International Conference on Communications Workshops (ICC), Budapest, Hungary, 9–13 June 2013; pp. 1057–1061.

37. Hafeez, K.A.; Zhao, L.; Ma, B.; Mark, J.W. Performance Analysis and Enhancement of the DSRC for VANET's Safety Applications. *IEEE Trans. Veh. Technol.* **2013**, *62*, 3069–3083.

Chapter 3

Embedding a Multi-agents Collaboration Mechanism into the Hybrid Middleware of an Intelligent Transportation System

Steven K.C. Lo and Huan-Chao Keh

1Department of Information Management, Jinwen University of Science and Technology, New Taipei City 231, R.O.C, Taiwan

2Department of CSIE, Tamaking University, New Taipei City 251, R.O.C, Taiwan

ABSTRACT

Even if wireless networks and mobile computing technologies have comprehensively developed in recent years. Letting people extract information anywhere at anytime is the goal of this development. But in the telematics domain, the drivers can obtain road information through radio or certain in-car equipment, there is still a wide gap with regard to the synchronization of this information with actual road conditions. In the absence of adequate information, drivers often react to conditions with behaviors that do not contribute to their own driving goals but rather cause more complicated traffic conditions. Therefore, this study employs a process known as multi-agent collaboration. Information exchanged between the features and established mutual communication and collaboration mechanisms is applied in Intelligent Transportation Systems (ITS). By allowing drivers to have distributed communication with other vehicles to share driving information, collect information and/or submit their own reasoned driving advice to other drivers, many traffic situations could effectively be improved and the efficiency of the computing processes could be improved through distributed communication. This study proposes an architecture design for middleware that includes vehicle information, navigation, announcements and communication which could prove to be a more convenient and efficient intelligent transportation system.

1. INTRODUCTION

With the significant developments in wireless networks and mobile computing technology in recent years, capturing the mobile sources and locations of data has become a very important information service. For this reason, we use multi-agents (Al-Sakran, 2006; Dastani and Gomez-Sanz, 2005; De Meo et al., 2007; Davidsson and Wernstedt, 2002; Horling et al., 2006; Kong et al., 2006; Lesser, 1999; Lo et al., 2009; Ottens et al., 2009; Rahimi et al., 2006; Tweedale et al., 2007; Zambonelli et al., 2003) communication, coordination mechanisms (Jennings et al., 1998; Liu and Yao, 2004; Zhu et al., 2010) and existing mobile agent technology to integrate and improve the system. We develop a set of basic (i.e., access point) operations for modular information systems and query their environment and mobile agent data to design the message transmission mechanism. In the multi-agent communication and coordination mechanism, we propose four different interactive modes. Based on these modes, we create transport mechanism messages that can primarily follow the movement of mobile agent locations, enabling the prompt and accurate sending and receiving of information. The message template includes space for a mobile agent. The message transmission mechanism consists of three main components: (1) central server (the control server) (2) access point (regional based station) and (3) mobile data sources (sources). The main components of the mechanism involves Single Agent Systems (SAS) and Multi-Agent Systems (MAS) (Ellis and Wainer, 1994; Ilarri et al., 2008; Kotz and Gray, 1999; Lo et al., 2006; Shah et al., 2009; Yau et al., 2003), as shown in Table 1. The related mechanism is designed to execute tasks in the background through an application that drives an event trigger.

According to studies on cognitive access Points, it is possible for mobile communication systems to set up a wireless base station. The message transmission (Deugo, 1999) principle is also designed to allow for adjustments to the application module settings which is called access architecture in this study. The mobile agent (Baek and Yeom, 2006; Hart et al., 2001) message on Point transport mechanism is informal in its application and principles.

The mobile agent applies this transmission mechanism in the wireless network which has the advantages of reducing servo host loading, increasing the quality of network-connectivity, effectively managing information and facilitating easy cross-platform communication. It also provides the message object with mobility, autonomy and scalability. In the future, we hope that through this action mechanism, an agent can be integrated into highly portable mobile devices (O'Hare et al., 2006; Ozen et al., 2004), such as notebook computers, Personal Digital Assistants (PDA) and mobile phones. Last, the agent's mobility and autonomy can be enhanced through this mechanism. Thus, no matter where we go, we can access the most immediate and accurate information.

In addition, the study uses a defined interaction communication mode with an improved message transmission mechanism for freeway information. In the absence of adequate information, drivers often react to conditions with behaviors that do not contribute to their own driving goals but rather cause more complicated traffic conditions. For example, drivers compete to overtake each other in traffic but after a while find that their vehicles are still surrounded by other vehicles, showing that overtaking does not contribute to effective forward motion. In a traffic jam (for example, if the speed of the vehicle in front is too slow or there has been a car accident), it would be helpful for drivers to know whether they should change from the freeway to another road or wait patiently because the traffic jam may be resolved quickly. Therefore, we use a multi-agent collaboration and message transmission mechanism so that drivers can distribute communication with other vehicles, share their driving information, collect information and/or submit their own reasoned driving advice to the drivers. In addition, we propose a hybrid architecture that integrates OBU and RU to assist and apply in the fields of intelligent transportation systems and VANET (Bernardos et al., 2007; Cencioni and Pietro, 2008; Hu et al., 2007; Isaac et al., 2008; Javanmard and Ashtiani, 2009; Kane et al., 2008; Lee et al., 2008; Santa et al., 2008; Zhou et al., 2009).

Table 1. Agent system classification

	Single-Agent System (SAS)		Multi-Agent System (MAS)	
Class	Local	Network	Central and distributed	Fixed and migrating
Application	Personal assistants meeting schedulers	Personal assistants smart mailboxes	Distributed problem solving	Telecommunication

2. ESTABLISHED MULTI-AGENTS COLLABORATION MECHANISM

This section mainly focuses on the features of multi-agent interactions. We define the existence patterns of their interactions. When an agent wants to communicate with other agents, it cannot go beyond the scope of its basic mode of communication. In general, we can discuss many different types of communication among agents, such as one-to-one, one-to-many or many-to-many interactions. We first explore the simple one-to-one model. This study develops an agent announcement state transition diagram which represents the fact that an agent wants to communicate or interact with another agent. If an agent wants to access information from others or wants to share its own information, it must declare this intention. The announcement state transition is shown in the Figure 1.

In Figure 1, there are five possible states: announcing, suspending, mutating, critical and accepting. These states represent the possible states of the announcement process if one agent wants to communicate with another agent. We now explain the significance of each state.

Announcing: Announcing is an initial state of an Agent. This state indicates that this agent can communicate with other agents and pass messages.

Suspending: When the agent completes its task or is no longer able to coordinate, it can return from the suspending state to the initial announcing state to wait. Mutating: After the implementation state of the agent, if there is any other declaration of implementation, we can return the state of the agent back to announcing through a mutating state and re-coordinate.

Critical: The agent enters this critical state through a sequential process of coordination and implementation.

Accepting: When the agent can only accept messages from other agents, then it enters this state.

A collaborative model of agents: In general, animals interact with each other in several different interactive modes. For instance, they cooperate with each other to survive, behave selfishly or compete. The interactive behavior could be the result of sharing a living space with others or competing with others for benefits and/or achievement of a common goal. In the multi-agent interaction model, there is a similar notion of synergy.

The purpose of this study is to identify the possible forms of interactive behavior between agents, namely, cooperation, altruism, selfishness and competition. Figure 2 shows the interactive behaviors of the source agent communicates with the destination agent.

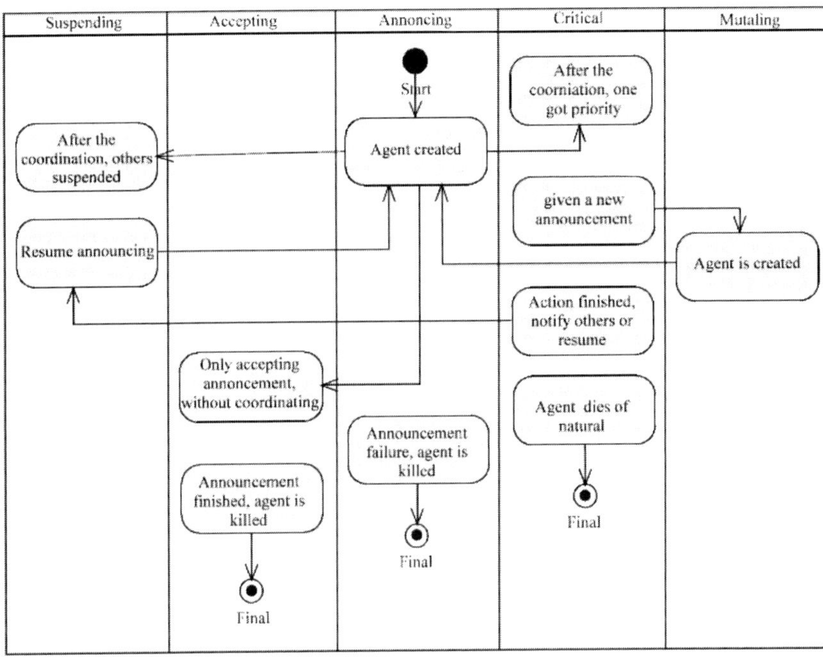

Figure 1. Agent announcement state transition

Cooperation: The communication and coordination between the two agents leads to win-win results. Such interaction is called cooperation. The cooperation mode of interaction between agents is shown in Figure 3. Based on the definition of cooperation, the source agent can specify a complex task or make a request for other agents to complete. This model can effectively reduce execution time and obtain efficient outcomes.

Description

- Source agent A strategically decides to send a demand (or request) to the Destination agent B and C and hopes to receive their responses

- Destination agent B strategically decides to send a demand (or request) to the Source agent A and Destination agent D and hopes to receive their responses
- When Destination agent B and C complete the task requested by Source agent A, Destination agent B and C send their responses to Source agent A
- Similarly, when Source agent A and Destination agent D complete the task requested by Destination agent B, Source agent A and Destination agent D send their responses to Destination agent B
- As Source agent A and Destination agent B respond to each other's demands, they share the processing results with each other. Thus, Source agent A and Destination agent B can achieve their co-operative goal together

Algorithm 1: Agent announcement state transition algorithm

```
Announce ( )
{if Destination Agent = free then
    if Check (Announce) = true then
    Reply (accept);
    else
    if Check (Announce) = not to conform to logic or not complete then
    Reply (repair);
    else
    Reply (error);
else
    Wait ( )
    end if;
    end if
    end if }
    Wait ( )
{if n <= times then n++;
    Reply (wait);
    Aunounce ( );
    else
    Reply (again);
end if }
```

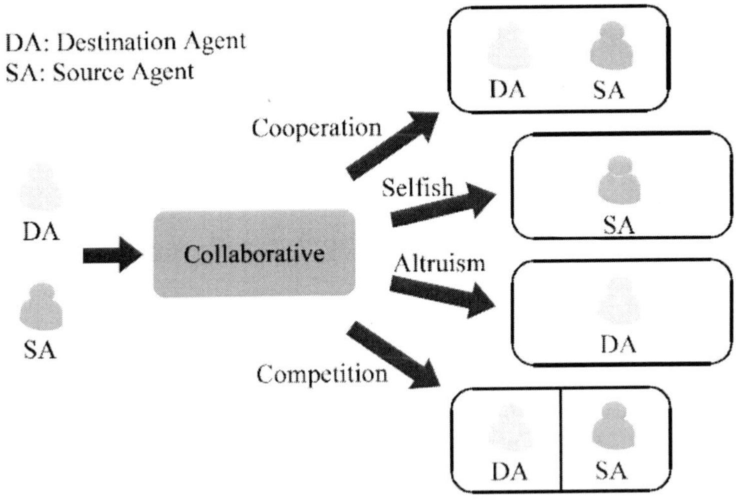

Figure 2. Multi-agent communication and collaboration

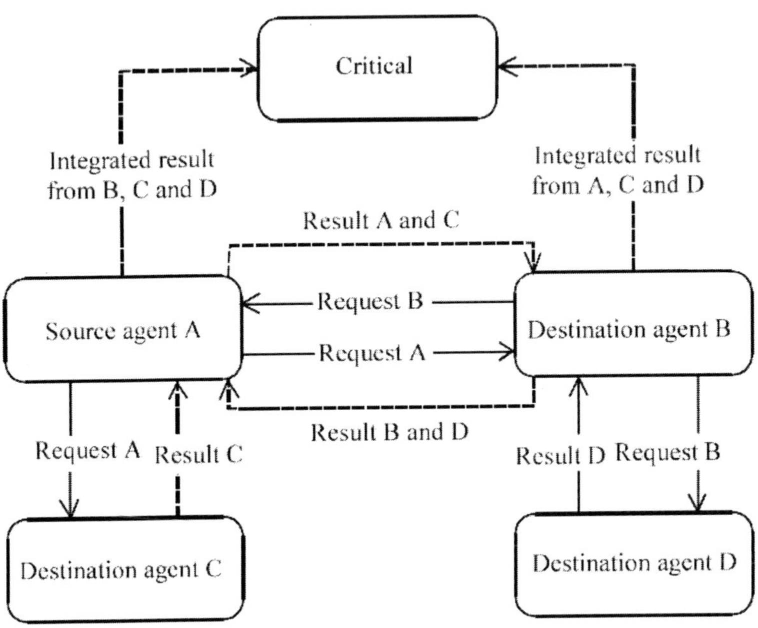

Figure 3. Cooperative behavior model

Algorithm 2: Cooperative behavior

```
SAA=new Agent; //Source Agent A
DAB=new Agent; //Destination Agent B
Result DACFunction (Agent Source A)
{DAC=new Agent;//Destination Agent C
rst=DAC.ExecuteTask (Source A);
Destroy (DAC);
Return (rst);}
Result SAAFunction (agent source B)
{ rst=SAA.ExecuteTask (Source B);
Return (rst);}
while (IsTaskFlag ==OK)
{   ResultC = DACFunction (SAA);
ResultA = SAAFunction (DAB);
SAA.Restult = ResultC+ResultA;                    }
Destroy (SAA);
```

Altruism: When two agents start communicating and coordinating, the Source Agent (SA) offers beneficial behavior to the Destination Agent (DA) but the Destination agent is not asked to do anything in return; the behavior of the Source agent is only beneficial to others. In this study, we call such communication Altruism. The altruism Interaction mode and the corresponding procedure between agents can be expressed as in Figure 4. The Source agent completes the related tasks and passes them to a particular agent or provides relevant information to help another agent successfully accomplish its objectives. Therefore, this particular agent can achieve his objectives through the altruism mechanism.

Description

- SA-A strategically decides to make a request to DA-C and to obtain its result
- DA-B strategically decides to make a request to SA-A and DA-D and to obtain its result
- When SA-A and DA D fulfill the request made by Agent B, SA-A and DA-D send their processing responses to DA-B

- As SA-A responds to the request made by DA-B, it will simultaneously share the processing results from DA-C such that DA-B benefits from the altruistic behavior of SA-A

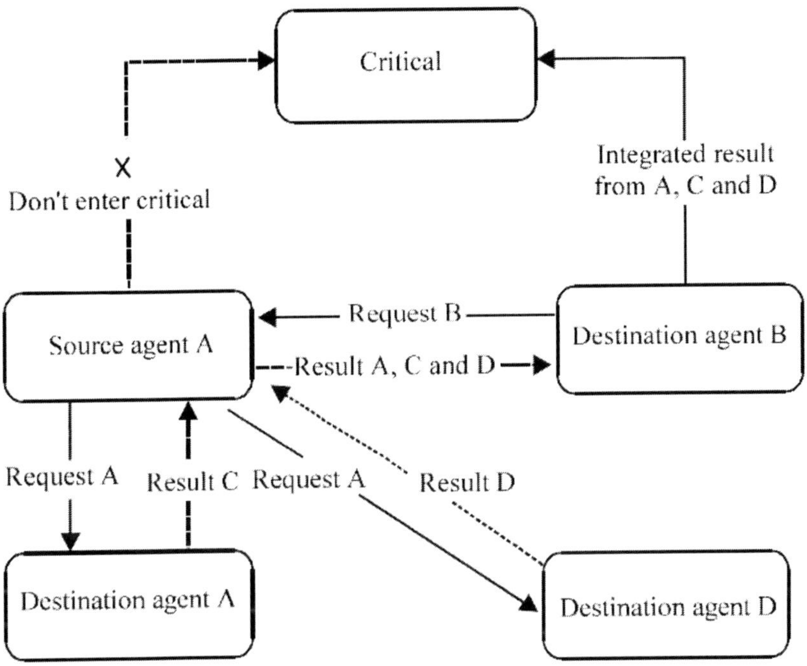

Figure 4. Altruism behavior model

Selfishness: When the two agents communicate and coordinate with each other, the Source agent may take selfish actions against the Destination agent. Such communication is labeled selfish. Figure 5 displays an agent showing the Selfish mode of communication. The given agent and other agents establish an initial stage of interaction. We assume that only a specific agent takes advantage of others and refuses to share its own resources and expertise with others.

Algorithm 3: Altruism behavior

```
SAA = new Agent; //Source Agent A
DAB = new Agent; //Destination Agent B
Result DACFunction (Agent Source)
{DAC = new Agent;//Destination Agent C
rst = DAC.ExecuteTask (Source );
Destroy (DAC);
Return (rst);}
Result DADFunction (Agent Source)
{DAD = new Agent;//Destination Agent D
rst = DAD.ExecuteTask (Source);
Destroy (DAD);
Return (rst);}
Result SAAFunction (Agent Source)
{ rst = SAA.ExecuteTask (Source);
Return (rst);}
while (IsTaskFlag = = OK)
{  ResultC = DACFunction (SAA);
ResultD = DADFunction (SAA);
ResultA = SAAFunction (DAB);
SAA.Restult = ResultC+ResultD+ResultA;            }
Destroy (SAA);
```

Description

- SA-A strategically decides to make a request to DA-B and DA-C and hopes to receive their processing results in return
- DA-A strategically decides to make a request to SA-A and DA-D and hopes to receive their processing results in return
- When DA-B and DA-C fulfill the request of SA-A, DA-B and DA-C send their processing responses to SA-A.
- DA-D completes the request from DA-B and sends its results back to DA-B but SA-A refuses to send the processing results from DA-C and its own resources to DA-B
- SA-A obtains resources from DA-C, obtains processing results from DA-D through DA-B and also gathers its own resources to achieve its goal

on time. However, DA-B is unable to obtain comprehensive information from SA-A; DA-B will not achieve its purpose in communication because of the selfish behavior of SA-A

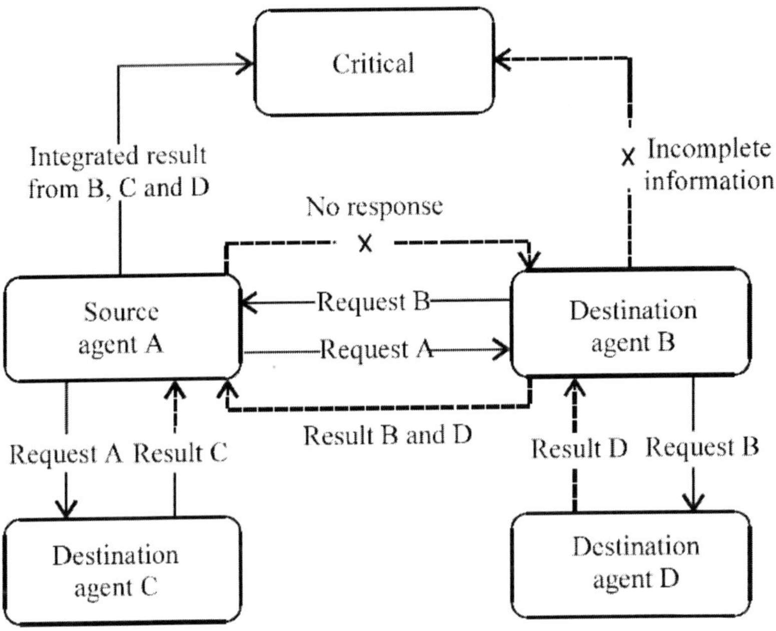

Figure 5. Selfish behavior model

Competition: When two agents communicate and coordinate with each other such that the action of the Source agent is not helpful to the destination agent and the Destination agent does the same in return, then the outcome is not conducive to either side; in this study, we define such communication as competition.

Figure 6 shows the interactions among agents in the competitive mode of communicating. The agent begins an initial stage with other agents. At this moment, the Source agent takes the advantage of a specific agent and does not share its own resources to further the best interests of others.

Algorithm 4: Selfish behavior

```
SAA = new Agent; //Source Agent A
Result DABFunction (Agent Source)
{DAB = new Agent; //Destination Agent B
rst = DAB.ExecuteTask (Source);
Destroy (DAB);
Return (rst);}
Result DACFunction (Agent Source)
{ DAC = new Agent; //Destination Agent C
rst = DAC.ExecuteTask (Source);
Destroy (DAC);
Return (rst);}
While (IsTaskFlag ==OK)
{  ResultB = DABFunction (DAB);
ResultC = DACFunction (DAC);
If (ResultB and ResultC)
OK;
else
FAIL;}
Destroy (SAA);
Destroy (SAB);
```

Description

- SA-A decides to send a request to DA-B, DA-C and DA-D. Furthermore, it also hopes to receive their results
- DA-B decides to send a request to SA-A, DA-C and DA-D. Furthermore, it also hopes to receive their results
- When DA-C and DA-D fulfill the request from SA-A, they (that is, DA-C and DA-D) send their processing results to SA-A. However, DA-B does not necessarily send its resources to the SA-A, depending on its own needs and competitive strategy
- When DA-C and DA-D fulfill the request from DA-B, they (that is, DA-C and DA-D) send their processing results to DA-B. However, SA-A does not necessarily send its resources to DA-B, depending on its own needs and competitive strategy

- SA-A and DA-B will respond partly, fully or not at all, depending on their mutual needs, to achieve their competitive goals
- SA-A and DA-B have been given resources from DA-C and DA-D but they lack detailed information about each other, causing them to engage in a series of competitive behaviors to achieve their goals if these goals cannot be integrated, both of them might be affected. One agent might grab or seek additional resources to be successful

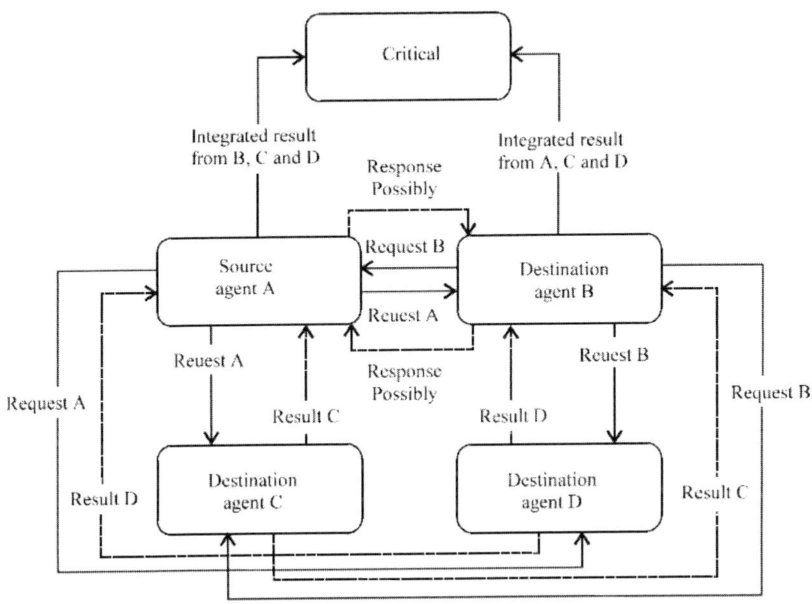

Figure 6. Competitive behavior model

3. IMPROVEMENT OF THE MULTI-AGENTS MESSAGE TRANSMISSION MECHANISM

Following the assessments of Dwight Deugo regarding message delivery mechanisms, it across five kinds of agents aimed at shortening the distance between the messaging and sending time, more reliable messages are provided as a guarantee to reduce the use of resources and construction costs. We present an improvement on the multi-agent message transmission mechanism.

Algorithm 5: Competitive behavior

```
SAA = new Agent; //Source Agent A
Result DABFunction (Agent Source)
{DAB = new Agent;//Destination Agent B
if (Source.goal != DAB.goal)
{ rst = DAB.ExecuteTask (Source);}
else
{ Break;}
Destroy (DAB);
Return (rst);}
Result SAAFunction (Agent Source)
{if (Source.goal != SAA.goal)
{ rst = SAA.ExecuteTask (Source);}
else
{ Break;}
Return (rst);}
While (IsTaskFlag ==OK)
{ ResultA=SAAFunction (DAB);
ResultB=DABFunction (SAA);                    }
Destroy (SAA);
```

Access point

Mobile data source transmission mechanism: As shown in Figure 7, at each node, data are generated and changed by the source based on the agent's movement which is called the Mobile Data Source (hereafter, MDS). Each MDS message is sent directly to the information jurisdiction area (or access point). If the agent is moving, the MDS changes the original data but we also need to update new information back to the access point which refreshes the storage and publishes the sources.

When the access point obtains information from the MDS, it retransmits it to the central server for comprehensive integration. The central server receives the incoming information from different parts of the access points and makes changes to the storage and then it broadcasts the updated information back to all access points. However, the information that is sent back is not the simple data that local access point needs but the most comprehensive information that could be requested. Therefore, the different parts of the access point must seek relevant information within their respective storage areas and release information to enable mobile agents to quickly access the data that they need.

Regional-based multi-agent message transmission mechanism-district mode: Through the access point MDS transmission mechanism described above, we can see the role of every access point and its accompanying features. We now explore these activities from the perspective of the mobile agents, specifically how they obtain the desired information at any given time and use it as a reference for their next step.

As shown in Figure 8, when the Mobile Agent (hereafter, MA) performs activities at each node, it is necessary to obtain valid reference information in the shortest possible time to prepare for the next step. Sometimes, the MA needs only certain region-specific information; then the MA needs only to send an inquiry signal through the access point in its precinct. Once the access point receives the MDS data, it will quickly transfer messages to the central server. After receiving all of the data from different access points, the central server will analyze and integrate the data and return updated, comprehensive data to different parts of the access point. The different parts of the access point will pick up the information from each area and place it in storage and then it can send an immediate reply to the MA.

If the MA moves to the next node, the data transmission from the access point will not be interrupted due to the movement of the MA. Because the transmitted information from the access point is continuous, as long as the receiving side of the MA is not closed, the transmission data will continue to come in until it reaches the jurisdiction of the next access point. At that moment, the original data will stop the transmission; thereafter, the MA will receive the information from the newly-located access point. The MA not only can obtain data from a single access point but they also can obtain very comprehensive information because all received MDS information will be passed to the central server for integration and release. If the MA does not want to know the information within a specific region but only wants to know the consistency of information among regions, then the MA can send an inquiry signal directly to the central server. When the central server receives all messages sent from each access point, it will integrate real-time, updated information and place it in storage so that it can reply immediately to the MA with the results.

As shown in Figure 9, when ma (1) wants to move to another area, originally it received the transmitted information from access point (1) but when ma (1) reaches the district from access point (2), then access point (1) will stop its data transmission. Thereafter, ma (1) receives transmitted information from access point (2). When ma (2) is moving, it is expected to be able to predict region-wide data as a reference for the next step. Ma (2) can send an inquiry directly to the central server so that the central server can integrate all incoming data from each access point. Therefore, the central server requires only real-time and comprehensive data to place in the storage; it could immediately reply to MA (2) with the requested data.

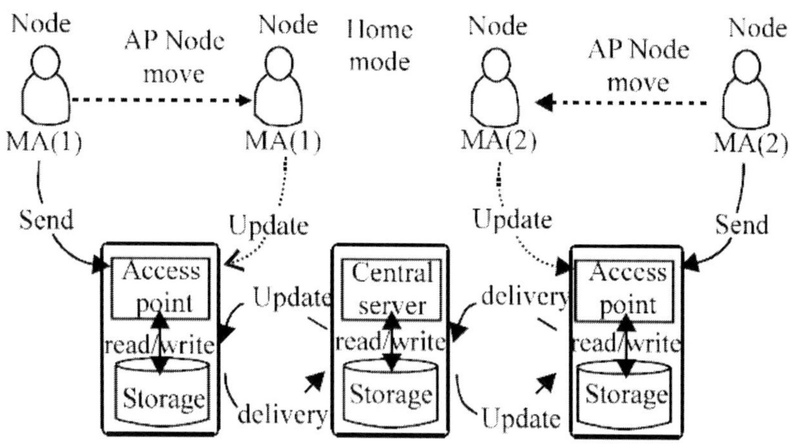

Figure 7. Access point action of sources delivery mechanism

Wide-area-based multi-agent message transmission mechanism-domain mode: The regionally-based access point message transmission mechanism is mainly applicable to the regional wireless communications infrastructure. MAs not only can obtain information within the specific region through the access point but also can apply directly to the central server to obtain the all information set. This message transmission process is considered to pose too high a risk of overloading the central server. If none of the MAs go through access points to access the information, then all of them will access the data through the central server, resulting in a slow rate of data transfer, incomplete information and even network failure. Therefore, to make the message transmission from access point to mobile agent more efficient and to support extensive use, it needs to be upgraded cross-regionally and be able to function either synchronously or asynchronously. Most MAs are able to receive the most immediate and most accurate information by providing their host with information to move more efficiently in traffic and facilitating more secure data transfer. Therefore, based on present analysis and research, we propose a messaging mechanism based on the advanced access point to mobile agent pathway to achieve the desired goals.

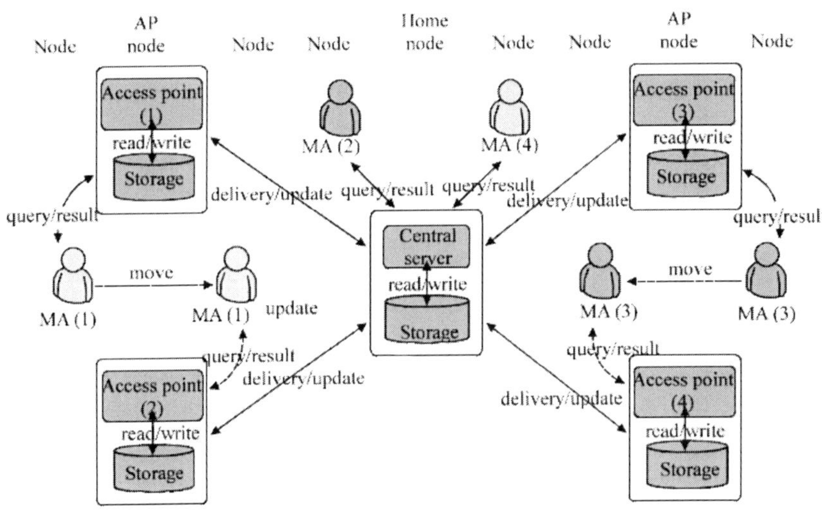

Figure 8. Message transmission mechanism in district mode: the overall transmission infrastructure

As shown in Figure 10, after the access point has received the message from MDs, it passes the information to the central server. The central server gathers all data from each access point, completes integration and analysis and then sends the comprehensive updated data back to each access point. There is one major difference from district mode: each access point no longer only places the data from the given region into storage but also places it in conjunction with the comprehensive updated information in storage.

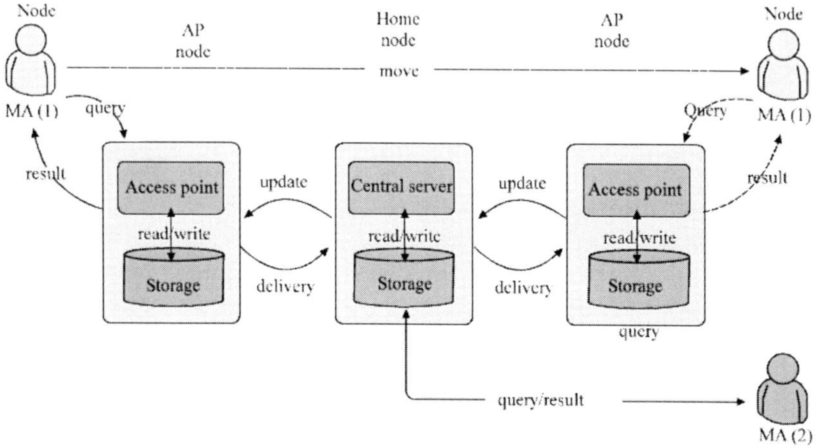

Figure 9. Message transmission mechanism in district mode: An agent delivery infrastructure

When the MA moves to another node, it can even estimate pre-treatment reference values based on the current node through the comprehensive information it receives from the access point. Of course, to avoid the ma directly grabbing information from the central server and to reduce the host loading traffic problem, all data are broadcast into the storage of each access point. This splits the data flow across every access point and allows them receive the data cooperatively. The MA is thus able to grasp more complete sources more efficiently in real time.

As shown in Figure 11, when MA (1) wants to move to another location, it can read data that have been updated, integrated and transferred from the central server and plan pre-emptive action. There are several mobile agents (i.e., MA (2) to (x)) within the framework path from access point (1) to (2). The comprehensive information requested by the MA which is provided by the central server, is deposited in the storage areas of access point (1) and (2). All MAs within the scope of this route can be assigned to access point (1) and (2) to receive data. In this way, it is possible to share the loading flow and control the entire network's traffic on each of the servo hosts. However, MA (2) through MA (x) are in the path range between access point (1) and (2). The information on this part of the path is provided exclusively from access point (1); thus, when MA (2) through MA (x) obtain a message from this route, it must be fully managed by access point (1).

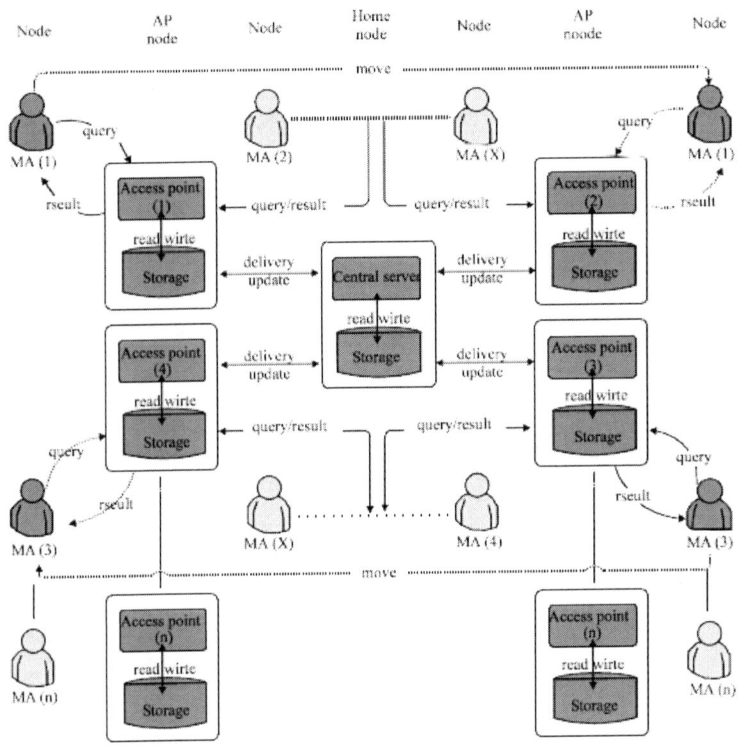

Figure 10. Message transmission mechanism in domain mode: the overall transmission infrastructure

4. THE MIDDLEWARE ARCHITECTURE BETWEEN VEHICLE AND INFRASTRUCTURE

We now draw on our definitions of the multi-agent coordination mechanism and message transmission models (that is, district mode and domain mode, respectively) discussed above. We propose a transportation middleware architecture between the application layer (i.e., the vehicle) and the network layer (i.e., the freeway infrastructure). In Figure 12, the application layer mainly represents the operation system on OBU. It shows the requests from the user and the results from the inference engine. The network layer represents related devices on RU, such as an access point or base station. It is composed of four parts, namely, the vehicle information manager, navigation manager, announcement manager and communication manager. We define them as follows.

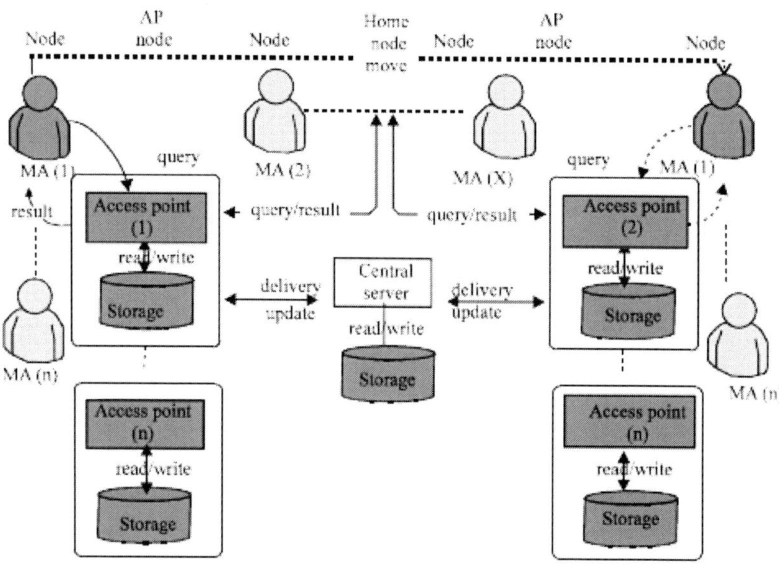

Figure 11. Message transmission mechanism in domain mode: an agent delivery infrastructure

Vehicle information manager: A vehicle Information agent can manage all information on a vehicle, including speed, destination, functions and so on. Moreover, it can record these data into a database and communicate with other managers.

Navigation manager: A vehicle positioning agent can process the dynamic position of vehicle. Then the Navigation Information agent can integrate the information and store it into the knowledge base. It also can provide them with data from a temporal/spatial inference engine to provide driving suggestions.

Announcement manager: This manager is responsible for communicating with other vehicles. It can send its own information to others for reference and it can also receive messages or information from other vehicles. When it receives information from others, it allows the navigation manager to evaluate it. So, this manager would play an important role in interacting with other vehicles. Moreover, when it wishes to announce information to other vehicle, it can follow the agent announcement state transition mechanism to communicate.

Communication manager: the communication manager is in charge of communication between other managers and network elements. When it wants to send a message to communicate, it can follow the district mode or domain mode. The central server information manager is equivalent to the core system management module. It is responsible for its own area access point and for processing data. In addition, the information manager is responsible for communicating internal messages with the mobile agent manager of the access point, including various query types. The administrator needs to send the best mobile agent based on the best data access strategy to process the task and to deposit the response into the database of the access point. Finally, if the system's data management strategy supports the updating of cache Data, then the information manager will be responsible for the maintenance and management of the cache on the access point.

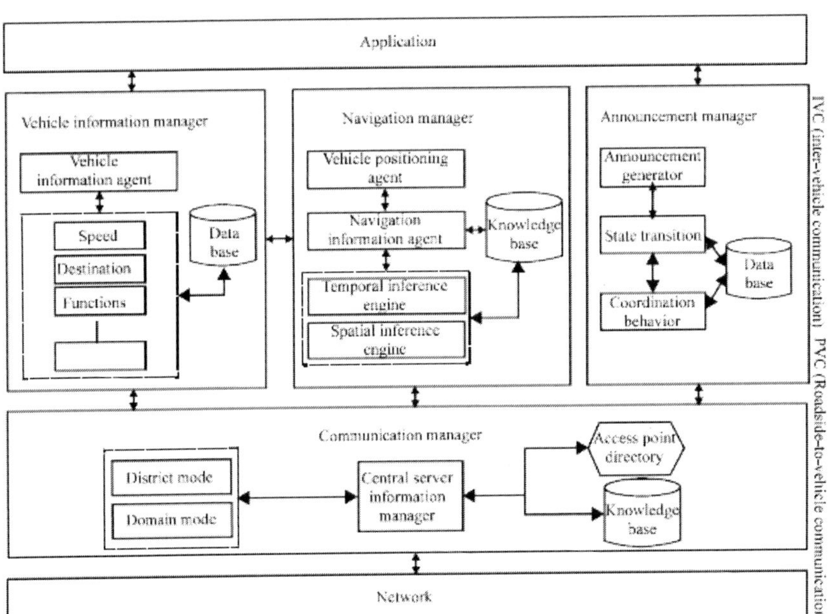

Figure 12. Overview of middleware architecture

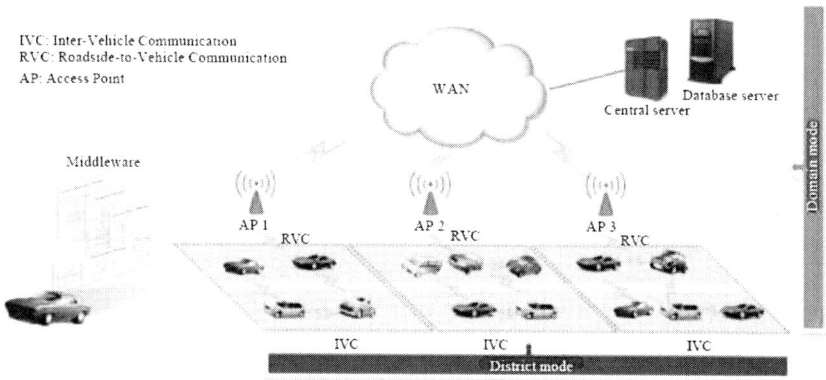

Figure 13. Integration of message transmission and middleware

In addition, the knowledge base is a repository with considerable capacity for data storage. The information manager is responsible for data management. If the data do not support the updated cache data management strategy and the long-distance data are changed, then the copy in the database should be deleted. However, when the cache data management strategy does support the updates, the information manager is responsible for maintaining the consistency of the database information.

As shown in Figure 12 and 13, based on these communication behaviors (cooperation, altruism, selfish and competition) each agent's information will be interactively shared and transmitted. This part of the study will be defined as the inter-vehicle communication (ivc) and it may facilitate regional management through the access point of each region. Access points for each region may be integrated through the central server with access points in other regions and thus share their useful regional data with other regions. We can use the message transmission mechanisms described in this study. We define as Roadside-to-Vehicle Communication (RVC).

In Figure 13, access point manages the defined cell area and all mobile data sources and stores the data from each operation in the directory. Via the wireless communication media for all mobile terminals, such as access points or base stations, it provides the wireless data for mobile terminals. From the client (such as obu) it must be accessed through the access point for data processing and communication transmission.

The central server information manager is the core system management module. It is responsible for its own area access point and for processing data. It is responsible for communicating the internal messages with the mobile agent of the access point which includes query types. The administrator needs to send the best mobile agent based on the best data access strategy to process the task and to deposit the response into the database of the access point. Finally, if the system's data management strategy supports the updating of cache data, then the central server will be responsible for the maintenance and management of the cache on the access point. Whether cars interact each other in every dotted line or they interact with access point and central server, they must follow the message transmission mechanism under the district mode and domain mode.

5. CONCLUSION

Based on the importance of current mobile information access needs, this study proposes multi-agent interaction involving collaborative models and a messaging communication mechanism which can be generated using a basic structure. This structure meets the requirements of wireless-based location information services through location-related data retrieval. This study presents a system framework and various types of dynamic data management systems to access data. The framework is more efficient than the traditional basic strategy and also less costly. We use a traffic information application on the freeway as an example of which vehicles can coordinate with each other and take advantage of regional and wide-area wireless environments to achieve the most effective dispersion of computation costs.

This study proposed the message transmission mechanism to aim at shortening the distance between the messaging and sending time, more reliable messages are provided as a guarantee to reduce the use of resources and construction costs and these mechanisms and framework could be used as the V2V, the vehicle to AP and AP to the central server communication and collaboration. Moreover, each AP will send the data back to the central server to reduce loading and latency effectively. This is also an expandable architecture and very novel solution.

REFERENCES

1. Al-Sakran, H., 2006. An agent-based architecture for developing e-learning system. Inform. Technol. J., 5: 121-127.

2. Baek, J.W. and H.Y. Yeom, 2006. A timed mobile agent planning approach for distributed information retrieval in dynamic network environments. Informat. Sci., 176: 3347-3378.

3. Bernardos, C.J., I. Soto, M. Calderon, F. Boavida and A. Azcorra, 2007. VARON: Vehicular Ad hoc route optimisation for NEMO. Comput. Communicat., 30: 1765-1784.

4. Cencioni, P. and R.D. Pietro, 2008. A mechanism to enforce privacy in vehicle-to-infrastructure communication. Comput. Communicat., 31: 2790-2802.

5. Dastani, M. and J.J. Gomez-Sanz, 2005. Programming multi-agent systems. Knowledge Eng. Rev., 20: 151-164.

6. Davidsson, P. and F. Wernstedt, 2002. A multi-agent system architecture for coordination of just-in-time production and distribution. Proceedings of the ACM Symposium on Applied Computing, (ASAC`02), ACM Press, pp: 294-299.

7. De Meo, P., A. Garro, G. Terracina and D. Ursino, 2007. Personalizing learning programs with X-Learn, an XML-based, user-device adaptive multi-agent system. Informat. Sci., 177: 1729-1770.

8. Deugo, D., 1999. Mobile agent messaging models. School of Computer Science, Carleton University.

9. Ellis, C. and J. Wainer, 1994. A conceptual model of groupware. Proceedings of the ACM Conference on Computer Supported Cooperative Work, (ACCSCW'94), USA., pp: 79-88.

10. Hart, D., M. Tudoreanu and E. Kraemer, 2001. Mobile agents for monitoring distributed systems. Proceedings of the 5th International Conference on Autonomous Agents, (ICAA'01), ACM Press, pp: 232-233.

11. Horling, B., V. Lesser, R. Vincent and T. Wagner, 2006. The soft real-time agent control architecture. Autonomous Agents Multi-Agent Syst., 12: 35-92.

12. Hu, F., A. Rughoonundon and R. Krishnaram, 2007. Using emulation instead of simulation to analyze synchronized higher layers performance in wireless networks. Inform. Technol. J., 6: 941-953.

13. Ilarri, S., E. Mena and A. Illarramendi, 2008. Using cooperative mobile agents to monitor distributed and dynamic environments. Inf. Sci., 178: 2105-2127.

14. Isaac, J.T., J.S. Camara, S. Zeadally and J.T. Marquez, 2008. A secure vehicle-to-roadside communication payment protocol in vehicular ad hoc networks. Comput. Commun., 31: 2478-2484.

15. Javanmard, A. and F. Ashtiani, 2009. Analytical evaluation of average delay and maximum stable throughput along a typical two-way street for vehicular ad hoc networks in sparse situations. Comput. Commun., 32: 1768-1780.

16. Jennings, N.R., K. Sycara and M. Wooldridge, 1998. Roadmap to agent research and development. Autonomous Agents Multi-Agent Syst., 1: 7-38.

17. Kane, L., B. Verma and S. Jain, 2008. Vehicle tracking in public transport domain and associated spatio-temporal query processing. Comput. Commun., 31: 2862-2869.

18. Kong, S.H., S.D. Noh and Y.G. Han, 2006. An agent-based collaborative assembly process planning system. Int. J. Adv. Manuf. Technol., 28: 176-183.

19. Kotz, D. and R.S. Gray, 1999. Mobile agents and the future of the Internet. ACM Operat. Syst. Rev., 33: 7-13.

20. Lee, W.H., S.S. Tseng and C.H. Wang, 2008. Design and implementation of electronic toll collection system based on vehicle positioning system techniques. Comput. Commun., 31: 2925-2933.

21. Lesser, V.R., 1999. Cooperative multiagent systems: A personal view of the state of the art. IEEE Trans. Knowledge Data Eng., 11: 133-142.

22. Liu, J. and C. Yao, 2004. Rational competition and cooperation in ubiquitous agent communities. Knowl. Based Syst., 17: 189-200.

23. Lo, S.K.C., H.C. Keh, Y.H. Lin and W. Jo-Chi, 2009. Integrated the intelligent agent behavior model and billing service into communication system. Inform. Technol. J., 8: 668-677.

24. Lo, S.K.C., J.C. Lin, H.C. Keh, S.J. Wu and Y.H. Lin, 2006. The intelligent agent of the billing service between the PSTN and VOIP. Inform. Technol. J., 5: 51-56.

25. O'Hare, G.M.P., M.J. O'Grady, C. Muldoon and J.F. Bradley, 2006. Embedded agents: A paradigm for mobile services. Int. J. Web Grid Serv., 2: 379-405.

26. Ottens, K., N. Hernandez, M. Gleizes and N. Aussenac-Gilles, 2009. A multi-agent system for dynamic ontologies. J. Logic Comput., 19: 831-858.

27. Ozen, B., O. Kilic, M. Altinel and A. Dogac, 2004. Highly personalized information delivery to mobile clients. Wireless Netw., 10: 665-683.

28. Rahimi, S., J. Bjursell, M. Paprzycki, M. Cobb and D. Ali, 2006. Performance evaluation of SDIAGENT, a multi-agent system for distributed fuzzy geospatial data conflation. Inform. Sci., 176: 1175-1189.

29. Santa, J., A.F. Gomez-Skarmeta and M. Sanchez-Artigas, 2008. Architecture and evaluation of a unified V2V and V2I communication system based on cellular networks. Comput. Commun., 31: 2850-2861.

30. Shah, N., R. Iqbal, A. James and K. Iqbal, 2009. Exception representation and management in open multi-agent systems. Inf. Sci., 179: 2555-2561.

31. Tweedale, J., N. Ichalkaranje, C. Sioutis, B. Jarvis, A. Consoli and G. Phillips-Wren, 2007. Innovations in multi-agent systems. J. Network Comput. Appli., 30: 1089-1115.

32. Yau, S.M.T., H.V. Leong and A. Si, 2003. Distributed agent environment: Application and performance. Inform. Sci., 154: 5-21.

33. Zambonelli, F., N.R. Jennings and M. Wooldridge, 2003. Developing multiagent systems: The Gaia methodology. ACM Trans. Softw. Eng. Methodol., 12: 317-370.

34. Zhou, L., G. Cui, H. Liu, D. Luo and Z. Wu, 2010. NPPB: A broadcast scheme in dense VANETs. Inform. Technol. J., 9: 247-256.

35. Zhu, C.Y., Y.Y. Du and P. Li, 2010. A routing-based dynamic workflow. Inform. Technol. J., 9: 561-568

Chapter 4

Development of a Real-Time Model of the Occupancy of Short-Term Parking Zones

Reinhard Hössinger[1], Peter Widhalm[2], Michael Ulm[2], Klaus Heimbuchner[3], Eike Wolf [4], Roland Apel[5], Tina Uhlmann[1]

[1]Institute for Transport Studies, BOKU, Vienna, Austria

[2]Austrian Institute of Technology, Vienna, Austria

[3]Heimbuchner consulting GmbH, Vienna, Austria

[4]Ing. Dr. iur. Eike Wolf, Vienna, Austria

[5]ITP Unternehmensberatung GmbH, Vienna, Austria

ABSTRACT

The lack of reliable information about the occupancy of on-street parking places in the inner districts of large cities causes an unnecessarily high amount of parking search traffic. Previous attempts to solve the problem rely on accurate information systems that navigate drivers to the next available parking space. They have failed due to high costs for roadside sensors and are also controversial, because they indirectly support car use. A service that informs road users early about the situation at the destination would support a shift towards public transport modes and reduce parking search traffic particularly at times of unexpectedly high parking pressure. The real-time occupancy model presented in this article does not aim to indicate the occupancy of single parking lots, but to provide a sufficiently accurate description and reliable prediction of the occupancy at the destination. Three existing real-time indicators were tested for their predictive power: (i) location data of mobile-phone parking; (ii) counts of car parks in the vicinity; and (iii) traffic flow volumes. Experimental results indicated that models using mobile-phone parking data have the potential to improve predictions of the occupancy rate by detecting exceptional deviations from the average day curve.

KEYWORDS

ITS; Telematics; Parking management; Short term parking; Mobile phone parking; Real-time occupancy model

1 INTRODUCTION

Driver assistance systems (DAS) became very popular in recent years. The automobile industry and academic researchers spend much effort to develop such systems [16]. They can detect certain driving situations and provide car drivers with real-time information on different aspects such as travel time [1] and parking facilities. The parking process can involve several annoying tasks: finding a free parking space, manoeuvring into a narrow space, finding a ticket machine or filling in a prepaid ticket etc. As a result, there is growing interest in parking assistance systems.

There are several techniques, which provide drivers with real-time information on the occupancy of car parks and guide them to the next free parking space. Counting systems like traditional induction loops and video detection [7] as well as secured wireless networks are available in car parks to record entering and existing vehicles. The degree of occupancy (or number of free spaces) is displayed on variable message signs (VMS) and drivers are guided to available parking facilities with directional signs [15]. A recent contribution is “smart parking”, which has been implemented in many European and Japanese cities to improve the speed and efficiency of locating, reserving and paying for parking [18]. But even inside a car park assistance may be required to find a free space. There are vision-based systems [13] and wireless sensor networks (WSN), which have been developed and tested, e.g., by Seong-eun et al. [20], Jaeseok et al. [10],

Tang et al. [23] and Gongjun et al. [8]. They inform drivers about free parking spaces and route them directly to a free place.

Many drivers would welcome analogous information services for on-street parking spaces. This task is more difficult, because entries, exits, and parking spaces are less well defined and inappropriate for counting. On-street parking vehicles can be detected with panoramic street images taken with different cameras. Hirahara and Ikeuchi [9] used a laser-ranger finder and a line-scan camera taking epipolar-plane images (EPIs) for this task. This method delivers only static snap-shots. Most attempts to deliver real-time information about parked vehicles on-street rely on roadside sensors. For example, Berenger Vianna et al. [3] as well as Basu and Little [2] developed a system with linked parking meters to find free spaces. Parking meters were equipped with infra-red sensors to identify free and occupied parking spaces; the parking meters were connected with each other via radio frequency transceivers. Free parking meters send their position to the network and passing cars can receive this information as well. BMW developed a similar system in the project 'stadtinfoköln' with about 140 connected parking meters in Cologne [14]. This information service is available for 19 German cities and can be used for free via internet.

Wired sensors networks are impractical to cover large areas that lack a suitable electrical infrastructure because of high deployment and maintenance costs [21]. An alternative option for detecting the occupancy of on-street parking spaces are wireless sensor networks with autonomous sensor nodes [7]. Most of these systems use ultrasonic sensors or induction loops. The devices communicate via radio frequency transmitters; the occupancy information can be accessed through mobile devices [12] or being displayed on VMS. Cenerario et al. [4] took a different path. They employed protocols that vehicles can use to share information by means of V2V communication; it could for instance be used to share information about available parking spaces. A similar community-based system is currently tested in Vienna [22]. It enables users to indicate (release) free on-street parking spaces and to find these spaces via a mobile app. This system would however require a high share of market penetration to provide a complete picture, which can only be achieved in the long run.

An accurate detection of on-street parking spaces relies invariably on area-wide infrastructure on street or within vehicles, which causes high costs for implementation and maintenance. Such systems are also controversially discussed, because they aim at navigating drivers as quickly as possible to the next free parking space, that way making car use more convenient. It is commonly accepted that car use should not be further encouraged, in particular not in the inner districts of large cities. This argues for an information system, which informs road users early about the current and forthcoming parking situation at the destination—preferably prior to departure, when there are still more options available to adapt the behaviour.

Many drivers are able to roughly estimate the availability of short-term parking spaces at the destination based on experience. However, situational influences such as holidays, building sites, and special events introduce variability to the parking pressure, what encourages people to try their luck. As a result, parking search traffic amounts to 25 % of total traffic in the inner districts of Vienna [6]. Rye et al. [19] emphasises that managing information about parking availability should be part of demand management. A more reliable forecast of the occupancy could support the use of alternative options (public transport, Park & Ride, garage at the destination) and reduce parking search traffic at the destination. The self-regulating effect could be reinforced by variable parking fees depending on the demand. Such schemes are already in operation, e.g. in Amsterdam [17].

The city of Vienna offers an SMS payment service for short-term parking fees, which is called "HANDY Parken". More than 40 % of the short-term parking tickets are paid by this means. The success of HANDY Parken in Vienna opens the way for a solution, which is independent from on-street sensors. Each electronic ticket generates information about the duration of the parking process and the approximate location of the vehicle via cell location. The approach is based on the assumption that these data provide a probabilistic indicator for the number of parked vehicles, thereby enabling an estimation of the total occupancy. A real-time readout procedure of position data of HANDY Parken does not exist yet, but has to be developed.

2 OBJECTIVES

The objective of the research project this article reports on was to develop a real-time model for the occupancy of short-term parking zones, which works without roadside infrastructure, but only uses existing real-time data sources (Figure 1). Three different kinds of data were tested for their predictive power:

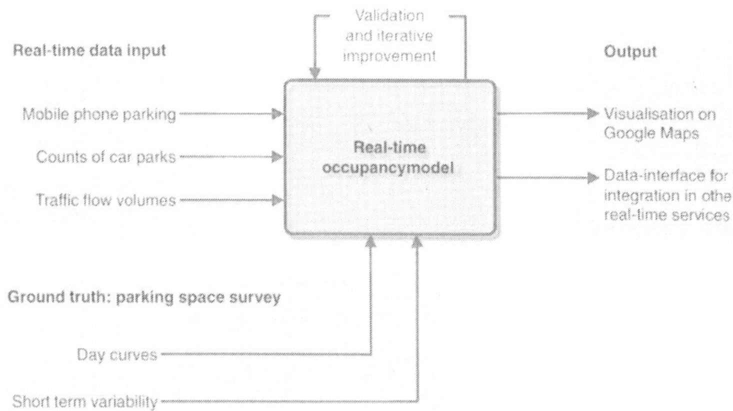

Figure 1. Data flow in the real-time occupancy model

- position data of short term parking customers, who buy an electronic parking ticket on their mobile phone via the HANDY Parken service;
- counts of short-term parking customers in car parks in or close to the target area;
- traffic flow volumes derived from around 300 automatic counting sites in Vienna.

We focussed on these data sources, because they are already available or could be gained at low cost. Moreover, this article is only about (i) testing for the predictive power of these data sources, (ii) developing a suitable model specification, and (iii) evaluating the performance of the model. The development of a physical model is not part of this article. The resulting model shall not exactly map the occupancy of single parking lots, but provide a sufficiently accurate description and reliable prediction of the parking pressure in the target area.

Vienna provides optimal conditions for this attempt due to the high share of mobile phone parkers. Parking meters are not available in Vienna and sending an SMS is apparently considered much more convenient than filling in a paper ticket. As a result, more than 40 % of parking tickets is paid by this service, what corresponds to more than a million SMS tickets every month. This makes HANDY Parken the most commonly used mobile service in Austria.

The mobile phone data are generated only during the validity period of the short-term parking zone and arrive just in time, which limits their usability for a forecast. Therefore, two further real-time data sources were tested in this study, which are available all time and may increase the reliability of forecasts. Both data sources are state of the art, but their use as real-time indicator of the occupancy of short-term parking zones remains to be proven.

The real-time occupancy model was developed and calibrated with data of an empirical parking space survey in two test areas. The most challenging tasks were to find out, (i) if the extraction of position data of HANDY Parken customers is technically and legally feasible and (ii) if the three aforementioned indicators can predict the real-time occupancy of the short term parking zones. The expected result is an online prototype, equipped with a visualisation on Google Maps and a data-interface for integration into existing travel information services. Preliminary results were presented at the ITS World Congress 2012 in Vienna.

3 REAL-TIME DATA SOURCES

The mobile phone parking data constitute the most promising indicator of the occupancy model. The HANDY Parken service has an agreement with several mobile network operators (MNOs), the largest of which participated in the research project. Reading out all SMS-tickets booked in a given time interval together with the permitted parking duration is technically and legally feasible. But, we also need to determine the position of the mobile phone from which the SMS was sent. It serves as proxy for the position of the car and is allocated to a particular street section. We aimed to perform this task by determining the coordinates of the active network cell. This procedure is less precise than GPS location, but can be universally applied, not only on smart phones. Due to the high density of network cells in the inner districts of Vienna, we expect a spatial accuracy of 100–200 m. Given that drives roam areas of a similar size when looking for a parking space, a more precise location is presumably not necessary for an information service about the occupancy of on-street parking spaces.

The Austrian Data Protection law under the rule of the European Data Protection law (EU-Dir. 95/46/EC and EU Dir. 2002/58/EC) [5] as well as the Austrian Telecommunication Act (§99 TKG2003) raise several cautious questions regarding privacy concerns: What kind of data retrieval is permitted for the MNO? At which stage of data processing will the aggregation take place? What level of aggregation is sufficient? These questions could not be answered in advance, but had to be discussed during the project. The data protection officers of the MNO finally decided against a data transfer despite an initial agreement. As a result, the HANDY Parken data could be made available for calibrating the model, but will not be available for operation.

The counts of short-term parkers in the car parks were easier to get, because all car parks in Vienna dispose of automatic counting systems. The counts are periodically retrieved and fed into the Viennese parking guidance and information system. However, for the development of an occupancy model it was necessary to distinguish between customers with long term contracts and short term customers. Since the parking information system requires only the total number of free parking spaces for its operation, this distinction is not common state of practice. We achieved an agreement with five of the six car park operators in the test areas. All of them were able to distinguish between long term and short term parkers and provided the counts during the parking space survey.

Traffic flow data could be collected without any difficulty. They are already collected at a central point, where they feed into a transport model, which is in service at www.AnachB.at. A data exchange protocol has been established, so that we could retrieve the data at any time.

4 PARKING SPACE SURVEY

The parking space survey conducted in the research project served two purposes: (i) to deliver the 'ground truth' for the calibration and validation of the occupancy model; and (ii) to deliver the day curves of those vehicles, which have a permanent parking permission, any other special permission, or no permission at all (fee dodgers). These parkers leave no electronic trace for real-time indication. The survey was conducted in two test areas as shown in Figure 3 with a total of 3,000 parking spaces. The areas vary in terms of residents-visitors ratio, supply of on- and off-street parking, spatial functions, and demographic structure. The survey was carried out in three periods in February, March and April 2012. It should be noted that the parking fee had been increased between period 1 and 2 from previously 1.20 to 2.00 €/h (the survey also served to estimate the effect of this intervention), so that only the 2nd and 3rd period are suitable for cross-validation under ceteris paribus conditions.

The survey started with a record of all parking supplies. For each available parking space it was indicated, whether it is a regular short term parking space, a special parking space for loading, disabled drivers, or any other specific group. The demand of parking spaces was afterwards observed periodically every 30 min from 8 am to 10 pm. Each parked vehicle was registered along with the type of parking permission in terms of permanent permission, fill-in ticket, or no visible permission. The latter could be fee dodgers or users of the mobile parking service. In order to distinguish between the two, we recorded the license plates of these vehicles and checked them against the HANDY Parken database, which allowed us to determine, which of these cars had a valid electronic ticket during the observation period.

5 OBSERVED OCCUPANCY RATES

The cars parking on short-term parking spaces can be grouped into the following categories according to the type of parking permission:

- permanent permission (mainly residents, but also businesses and other specific groups);
- short term permission in terms of a prepaid fill-in ticket;
- short term permission in terms of an electronic ticket; and
- no permission; these are fee dodgers who forgot to buy a ticket or deliberately try their luck that they are not enforced and punished; the share is 12 % in the total sample; it matches well with the experience of the parking surveillance authority.

The repeated observations of parking cars (every 30 min from 8 am to 10 pm on 11 days) result in 28 × 11 = 308 data points in time for each street section. The objective of the occupancy model is to predict the occupancy rate of an area rather than the number of parked cars. We thus translated the car counts into fractions of available parking spaces occupied by different categories as shown in Figure 3. The result reveals strong differences between the areas (Figure 2). The share of short-term parkers is approx. 40 % in the city centre (area 1), whereas it is only 20 % in the residential area (area 2). The day curves are also different. The occupancy in the city centre is highest during the day and again during the opening hours of pubs in the evening, whereas the residential area shows the highest occupancy in the morning before the residents leave and in the evening after they are back.

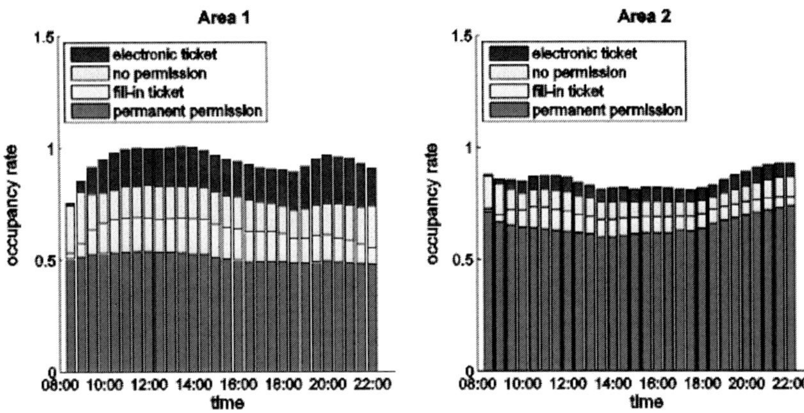

Figure 2. Share between different types of parking permissions as fractions of the estimated number of available parking spaces in 1st district (area 1) and 6th district (area 2)

6 ANALYSIS OF RELATIONSHIPS BETWEEN OCCUPANCY RATES AND REAL-TIME DATA SOURCES

This section includes the tests of linear relationships between the three kinds of real-time data and the occupancy of short term parking zones. The tests were performed using correlation coefficients. The results and conclusions presented in the following hold only true for Vienna. We do not have data outside Vienna, so we cannot discuss the applicability for other cities.

6.1 Counts of Short Term Parking Customers in Car Parks

Car park counts were collected from six car parks in or close to area 1 and three car parks in or close to area 2 (green triangles in Figure 3). We computed several correlations of car park data with parking space occupancy. As an illustration, Figure 3shows the correlations of the inflow of short term parking customers into the car parks with the average parking occupancy of the corresponding street sections. We found no systematic relationship between car park inflow and parking space occupancy, so that the former are no suitable indicator of the latter.

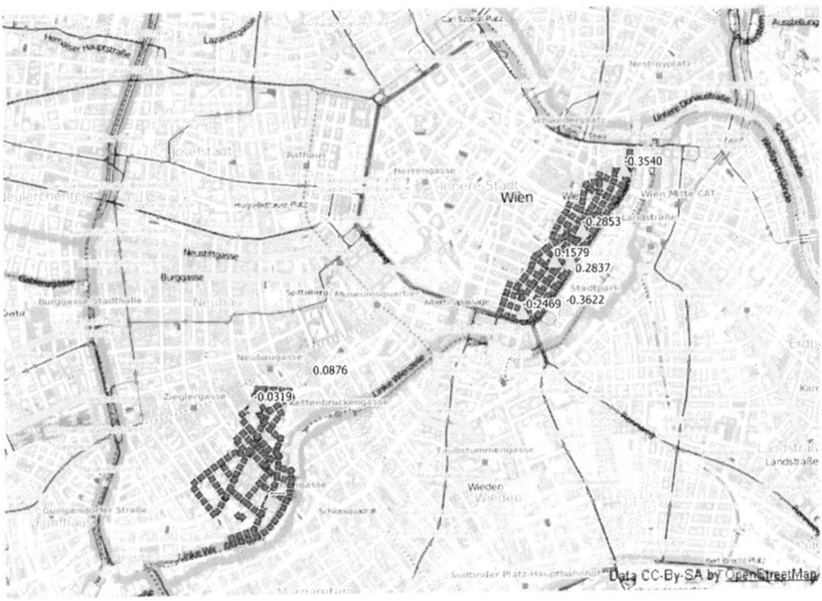

Figure 3. Correlation coefficients of garage inflow with mean parking occupancy in the corresponding area (*right top*: 1st district; *left bottom*: 6th district); red points = street sections, green triangles = car parks, numbers = correlation between inflow of short term parkers into car parks and on-street parking occupancy

Most of the car park operators, who participated in the advisory board of the research project and provided the counts of car park customers, held the view that the garage inflow is just the overflow of on-street parking demand. This view is not confirmed at all by the survey data.

6.2 Traffic Flow Volumes

We examined traffic flow data of 830 major roads in the Vienna region. They were derived from a traffic model operated by the provider of real time traffic information in the Vienna reagion [24]. The Model uses the data of some 300 automatic counting sites as input ([11]). In order to test for the predictive power of the traffic flows, we computed correlation coefficients with mean occupancy rates in both target areas. The results are given in Figure 4. Although we found some large correlations among the numerous candidates, they have different signs and don't reveal any recognisable geographical pattern. We thus considered the relationship between traffic flow data and parking space occupancy too fragile to be used as predictor in the real-time occupancy model.

Figure 4. Correlation coefficients of traffic flow data with mean parking occupancy rates in area 1 (1st district; *left graph*) and in area 2 (6th district; *right graph*)

6.3 Mobile Phone Parking Data

These data were the most promising candidates for indicating the occupancy of short term parking zones in real time. Our hypothesis was that (i) mobile phone parkers are a congruent subset of all short term parkers, and that (ii) the latter can be added to the number of permanent parkers, which evolves according to its own day curve independent from short term parkers. The mobile network operator involved in the research project refused to provide the location data of mobile phone parkers for privacy concerns (see Section 3). For the vehicles observed in the parking space survey, however, we could retrieve this information from the HANDY Parken database.

Table 1 shows the correlation coefficients of the number of electronic tickets with other kinds of permissions and with the total occupancy rate; it sheds some light on the relationships between these quantities. The strong positive correlation between electronic tickets and fill-in tickets indicates that mobile phone parkers are indeed a congruent subset of all short term parkers (hypothesis 1). However, the assumption of independence between short term parkers and permanent parkers (hypothesis 2) was rejected. The negative correlation with permanent permissions and fee dodgers (without permission) indicates a displacement effect, which may occur in cases when the total demand exceeds parking space supply.

Table 1. Correlation coefficients (*r*) and significance (*p*) between electronic parking tickets and other parking permissions in 1st district (area 1) and 6th district (area 2)

Kind of permission	Area 1		Area 2	
	r	*p*	*r*	*p*
Fill-in tickets	0.515	0.000	0.503	0.000
Permanent permissions	−0.259	0.000	−0.269	0.000
Cars without permission	−0.372	0.000	−0.418	0.000
All other permissions	−0.067	0.241	−0.238	0.000
Total occupancy rate	0.659	0.000	0.115	0.043

This pattern appears similar in both areas. A difference between the areas however appears when it comes to the correlation between electronic tickets and total occupancy rate. Area 1 shows a high positive correlation, whereas area 2 shows a low correlation. The reason is the lower share of permanent parkers in area 1. It causes that the relationship between electronic tickets and total occupancy rate is less adversely affected than in area 2.

Figure 5 shows a linear regression $\widehat{o} = kx + d$ of the total occupancy rate $\widehat{o}$ (last row in Table 1) on the percentage *x* of parking space occupied by cars with electronic tickets. The model explains $r^2 = 43.5$ % of the occupancy rate variance in area 1 and only $r^2 = 1.3$ % of the variance in area 2.

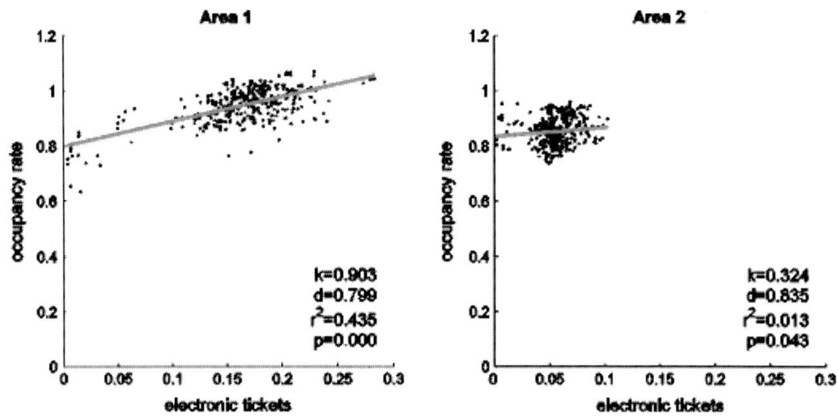

Figure 5. Linear regression of the total occupancy rate on the percentage of parking space occupied by mobile phone parkers in 1st district (area 1) and 6th district (area 2)

7 DEVELOPMENT OF THE REAL-TIME MODEL

In order to measure the improvement in prediction gained from the real-time information we defined two reference models, which differ in the underlying assumption:

1. The constant average model is based on the naive assumption that the occupancy rate stays at a constant average level at any time; it neglects the variability throughout the day. The expected average occupancy rate is simply estimated by

$$\overline{o} = \frac{1}{DT} \sum_d \sum_t o_{t,d} \tag{1}$$

2. over all days d and time intervals t from a sample of D days each divided into T time intervals. The Root Mean Square Error (RMSE) of this predictor is given by the sample standard deviation

$$s_o = \sqrt{\frac{1}{DT} \sum_d \sum_t \left(o_{t,d} - \overline{o}\right)^2} \tag{2}$$

3. The average day curve model is based on the slightly more advanced assumption that the occupancy rate follows the same average day curve on all days. The expected occupancy rate in time interval t on any day is estimated by

$$\overline{o}_t = \frac{1}{D} \sum_d o_{t,d} \tag{3}$$

where $o_{t,d}$ is the occupancy rate on day d in time interval t. This model explains 65.8 % of the variance of the occupancy rate in area 1 and 54 % in area 2, respectively. The resulting day curves are shown in Figure 2.

We tested the increase of predictive power gained from the real-time information against these models. The car park counts and traffic flow data were excluded from the outset, because no systematic relationship could be identified between these quantities and the parking space occupancy rate (see chapter 6); only the number of purchased electronic tickets was considered in the model. Predictions based on this indicator explain 43.5 % of occupancy rate variance in area 1 and 1.3 % in area 2 (Figure 5). This is still inferior to predictions based on the average day curve model, which account for 65.8 % and 54 %, respectively (eq. 3). Nonetheless, it is possible that the real-time indicator improves the predictive power, if added to the average day curve model. We tested this possible improvement in three respects:

- improvement of approximation of the occupancy rate in the current time interval;
- improvement of prediction of the occupancy rate of future time intervals;
- improvement of prediction of the occupancy rate in exceptional situations, in which the occupancy rate differs considerably from the usual day curve.

7.1 Approximation of the Occupancy Rate in the Current Time Interval

In order to determine the additional variance explained by the real-time indicator, we calculated the residuals of the average day curve model $\widetilde{o}_{t,d} = o_{t,d} - \overline{o}_t$ and, analogously, the residuals of the number of purchased electronic tickets $\widetilde{p}_{t,d} = p_{t,d} - \overline{p}_t$. The additional percentage of occupancy rate variance explained by a linear regression model

$$\widehat{o}_{t,d} = k\widetilde{p}_{t,d} + d \quad (4)$$

is $r^2 = 29.3\,\% \times (1 - 65.8\,\%) = 10\,\%$ in area 1 and $r^2 = 3.6\,\% \times (1 - 54\,\%) = 1.7\,\%$ in area 2, respectively. Figure 6 shows a plot of $\widetilde{p}_{t,d}$ against $\widetilde{o}_{t,d}$ for both areas along with the linear regression model according to eq. (4).

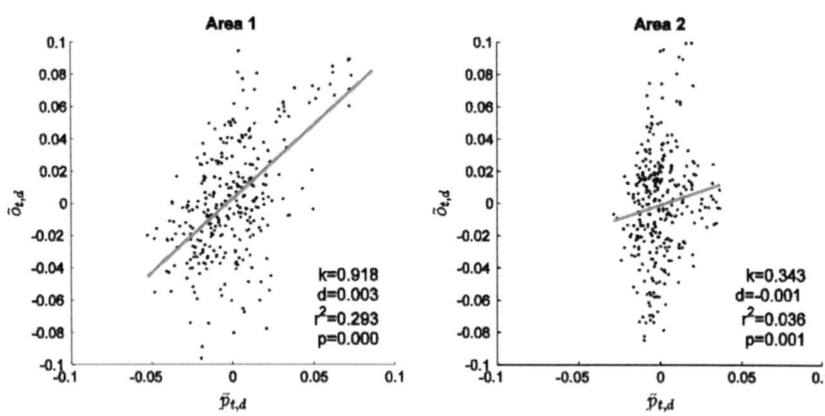

Figure 6. Linear regression of the residuals of occupancy rate $\widetilde{o}_{t,d}$ on the residuals of the number of mobile phone parkers $\widetilde{p}_{t,d}$ in 1st district (area 1) and 6th district (area 2)

7.2 Prediction of the Occupancy Rate of Future Time Intervals

The attempt to improve the prediction of the occupancy rate of a time interval in the near future by means of the number of purchased electronic tickets in previous time intervals is based on a linear model of the day curve of electronic tickets. The model uses the mean values p¯¯¯tp¯t and the covariance matrix **S** of electronic tickets purchased in previous time intervals to compute the predictor

$$\widehat{p}_{t+i} = \overline{p}_{t+i} + \mathbf{S}_{t+i;t-j,\dots,t}\mathbf{S}^{-1}_{t-j,\dots,t;t-j,\dots,t}\left(p_{t-j,\dots,t} - \overline{p}_{t-j,\dots,t}\right) \tag{5}$$

of the number of electronic tickets, which will be purchased in time interval *t* + *i*, given the number of electronic tickets purchased in time intervals *t* – *j*,…,*t*. The predictor of the occupancy rate in time interval *t* + *i* is again computed by a linear model, which uses the predictor of the number of electronic tickets pˆt+ip^t+i as predictor:

$$\widehat{o}_{t+i} = \overline{o}_{t+i} + k\left(\widehat{p}_{t+i} - \overline{p}_{t+i}\right) + d \tag{6}$$

The model parameters $\mathbf{S}$, $\overline{p}_t$, $\overline{o}_t$, k and d are estimated from empirical data.

A separate model was developed for both areas and subject to three cross-validation rounds. Within each round, two of the three survey weeks served for calibration, while the remaining week served for validation. In the prediction model given by eq. (6) we chose *j* = 0, i.e., only the electronic tickets purchased in the current time interval were used for prediction. Figure 7 shows the Root Mean Square Errors (RMSE) of occupancy rate predictions calculated 0, 30, 60 and 90 min ahead. Each chart shows a separate cross-validation round, which corresponds to one of the three weeks. The red line indicates the RMSE of predictions based on the average day curve model $\widehat{o}_{t+i} = \overline{o}_{t+i}$. The real-time indicator could in most cases not reduce the prediction error of the average day curve model.

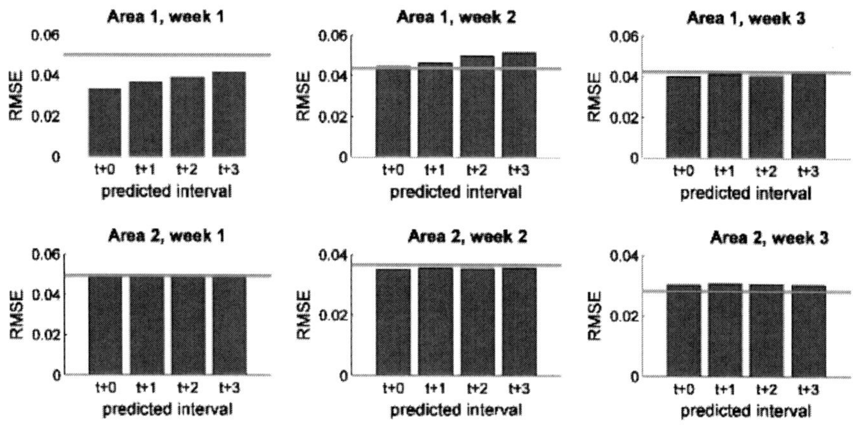

Figure 7. RMSE of a cross validation of occupancy rate predictions based on an average day curve model without (*red line*) and with (*blue bars*) real-time indication of purchased electronic tickets for time intervals t + 0 to t + 3

A direction for future work would be to examine the prediction accuracy of an implementation in a Kalman filtering framework with measurement model $p_t^* = h o_t^* + \epsilon_t$ and process model $o_{t+1}^* = \mathbf{F}_t o_t^* + \eta_{t+1}$ with observations $p_t^* = p_t - \bar{p}_t$, hidden state variables $o_t^* = (o_t - \bar{o}_t)$, observation model $h = k^{-1}$, observation noise $\epsilon_t \sim N(0, \sigma_t)$, state transition model $\mathbf{F}_t$ = cov $(o_{t+1}, o_t)/\mathrm{var}(o_t)$ and process noise $\eta_{t+1} \sim N(0, \xi_t)$.

7.3 Prediction of the Occupancy Rate in Exceptional Situations

Exceptional situations are by definition rare and will hardly come up numerously in three weeks survey period. We identified only one event, in which the occupancy rate differed considerably from the usual level. It was Shrove Tuesday in the 1st week, an event at which particularly many people visit the bars and pubs in the inner district of Vienna (area 1). This situation is included as one of three days in the upper left chart of Figure 7. In this particular case it turns out that the real-time model predicts the occupancy rate clearly better than the average day curve model. The error improvement is strongest in the current time interval, but it still appears in predictions of future time intervals. This result indicates that the real-time information about electronic parking tickets might increase the accuracy of an occupancy model in situations, in which the demand of parking place deviates from the usual level.

8 CONCLUSIONS AND OUTLOOK TO MODEL APPLICATION

We developed and tested a real-time occupancy model of short-term parking zones, which works without roadside infrastructure, but only uses existing real-time data sources. The following lessons we have learned from our work:

- Among the tested real-time data sources (electronic parking tickets, counts of car parks, traffic flow data) only the number and location of electronic tickets is able to indicate the occupancy rate in a given time interval and street section. However, this indicator is not yet available due to privacy concerns of the major mobile network operator in Austria.
- The occupancy rate of on-street parking spaces follows a strongly recurrent pattern on different days. An average day curve model thus predicts the occupancy at a given time interval very well. For the usual day-to-day variations of the occupancy rate it can hardly be outperformed by a real-time model.
- Unusual deviations from the day curve due to exceptional events are better predicted (and can only be predicted) by the real-time model. We had only one exceptional event in our three weeks survey, but such events will occur more often in the long run. There are various factors that significantly influence human behavior : holidays, day characteristics such as working day, weekend, day before or after holiday or weekend, big events, etc. This makes such a model very useful. Its further development requires data from longer periods. It is not necessary to anticipate the exceptional events; they would simply appear in the real-time signal, so that the model can learn from its own history for future events.

The unavailability of the location data of electronic parking tickets is not necessarily the end of the story. The negative decision was based on an extremely narrow interpretation of the legal regulations, so that we see a chance for a positive decision in a second attempt. A significant improvement can also be expected from the new standards for Intelligent Transport Services (ITS) and their subsection Cooperative IT, which are currently developed by the standard organisations ISO, CEN, and ETSI. The new standards will start at 2016 with the availability of new cars. The solution will be a combination of GPS-data, vehicle to vehicle (V2V) and vehicle to infrastructure (V2I and I2V) communication. They will deliver location data in the range of five meters. It can be used as real-time indicator of parked vehicles with unreached accuracy (among many other things).

A simple option to make the information service available to the users would be a stand-alone webpage as shown in Figure 8, where the user can chose the area and time for which the occupancy information shall be provided. The main goal is however to integrate the service into existing information services, e.g.

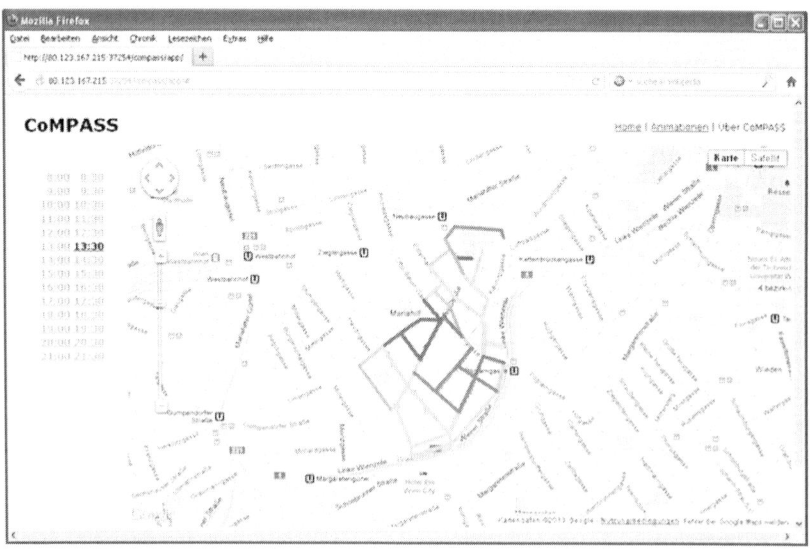

Figure 8. Internet information tool based on the real-time model

- the intermodal real-time traffic service AnachB.at;
- the mobile payment service HANDY Parken;
- the Viennese parking guidance system;
- traffic information services of broadcasting stations.

The end users' benefit of the information service is in either case a better knowledge of the current and forthcoming occupancy of on-street parking spaces at the destination. It could also include recommendations for alternative options in case of high parking pressure such as off-street parking facilities, intermodal transport, or public transport.

ACKNOWLEDGMENTS

Funding for this project is provided by the Federal Ministry for Transport, Innovation and Technology represented by the Austrian Research Promotion Agency (FFG).

REFERENCES

1. Ando, R., Mimura, Y.: A study to develop an information providing system on travel time. Int. J. Intell. Transp. Syst. Res. **8**(2), 77–84 (2010)
2. Basu, P., Little, T.D.C.: Wireless Ad Hoc Discovery of Parking Meters, Multimedia Communications Laboratory (MCL) Technical Report No. 01-08-2004 (2004)
3. Berenger Vianna, M.M., Da Silva Portugal, L., Balassiano, R.: Intelligent transportation systems and parking management: implementation potential in a Brazilian city. CITIES **21**(2), 137–148 (2004)
4. Cenerario, N., Delot, T., Ilarri, S.: A content-based dissemination protocol for VANETs: exploiting the encounter probability. IEEE Trans. Intell. Transp. Syst. **12**(3), 771–782 (2011)
5. Directive 95/46/EC of the European Parliament and of the Council of 24 October 1995 on the protection of individuals with regard to the processing of personal data and on the free movement of such data.

6. Dorner, A., Herry, M., Schuster, M.: Parkraumbewirtschaftung in Wien (Parking space management in Vienna). Werkstattbericht der Stadtplanung Wien, Magistratsabteilung 18, Vienna, Austria (1997) Federal Highway Administration: Advanced Parking Management Systems: A Cross-Cutting Study. U.S. Department of Transportation, Washington (2007)

7. Gongjun, Y., Weiming, Y., Rawat, D.B., Olariu, S.: SmartParking: a secure and intelligent parking system. IEEE Intell. Transp. Syst. Mag. **3**(1), 18–30 (2011)

8. Hirahara, K., Ikeuchi, K.: *Detection of street-parking vehicles from panoramic street image*, Proceedings Intelligent Transportation Systems, IEEE, 12–15 October 2003, **2**:993–998 (2003)

9. Jaeseok, K., Haeyong K., Jongsoo, J., Yongbin, S., Pyeongsoo M.: Field Deployment of a Large-Scale WSN for Parking Management System, Sensor Mesh and Ad Hoc Communications and Networks (SECON), 7th Annual IEEE Communications Society Conference, 21–25 June 2010, pp. 1–3 (2010)

10. Käfer, A., Fürst, B., Pediaditis, E., Gabert, K.: Straßenverkehrszählung Wien 2010. Auswertung Gemeindestraße A + B (Road traffic census Vienna 2010. Analysis of communal roads A + B). Magistrat der Stadt Wien, Magistratsabteilung 18, Vienna, Austria (2001)

11. Markoff, J.: Can't Find a Parking Spot? Check Smartphone, Online: http://www.nytimes.com/2008/07/12/business/12newpark.html?_r=2&ref=business&oref=slogin&oref=slogin (10.01.2012) (2008)

12. Micheloni, C., Foresti, G.L., Snidaro, L.: A network of co-operative cameras for visual surveillance. Vision Image Signal Proc. **152**(2), 205–212 (2005)

13. Motor-Talk: Nie mehr Parkplatzsuche: Persönlichen Stellplatz online reservieren, (Never searching a parking space again: reserve a parking space online), Online: http://www.motor-talk.de/news/nie-mehr-parkplatzsuche-persoenlichen-stellplatz-online-reservieren-t18460.html (10.01.2012) (2002)

14. Orski, K.: Best space scenario. Traffic Technol. Int. **2003**, 54–56 (2003)

15. Pellecchia, A., Igel, C., Edelbrunner, J., Schoner, G.: Making driver modeling attractive. IEEE Intell. Syst.**20**(2), 8–12 (2005)

16. Riedel, R.: Parkraumbewirtschaftung in Amsterdam NL (Parking Space Management in Amsterdam NL), Stadtplanung Wien, Magistratsabteilung 18—Stadtentwicklung und Stadtplanung (2006)

17. Rodier, C., Shaheen, S., Eaken, A. M.: Transit-Based Smart Parking in the San Francisco Bay Area, California – Assessment of User Demand and Behavioral Effects. Transportation Research Record: Journal of the Transportation Research Board, No. 1927, pp. 167–173. Transportation Research Board of the Acadamies, Wanshington, D.C. (2005)

18. Rye, T., Hunton, K., Ison, S., Kocak, N.: The role of market research and consultation in developing parking policy. Transp. Policy **15**(6), 387–394 (2008)

19. Seong-eun, Y., Poh, K. C., Taehong, K., Jonggu, K., Daeyoung, K., Changsub, S., Kyungbok, S., Byungtae, J.: PGS: Parking Guidance System Based on Wireless Sensor Network, pp. 218–222. Wireless Pervasive Computing, ISWPC 2008, 3rd International Symposium, 7–9 May 2008 (2008).

20. Sifuentes, E., Casas, O., Pallas-Areny, R.: Wireless magnetic sensor node for vehicle detection with optical wake-up. IEEE Sensors J **11**(8), 1669–1676 (2011)

21. Sofortparken.at: http://www.sofortparken.at; retrieved at March 08 2013 (2013)

22. Tang, V.W.S., Zheng. Y., Cao, J.: An Intelligent Car Park Management System Based on Wireless Sensor Networks, pp. 65–70. Pervasive Computing and Applications, 1st International Symposium, 3–5 August 2006 (2006)

23. Unterluggauer, A.: AnachB - die neue Live-Verkehrsinformation der Vienna Region (AnachB - The New Live Travel Information in the Vienna Region). Paper presented at the AGIT 2009 Symposium und Fachmesse für Angewandte Geoinformatik, Salzburg, Austria (2009)

Chapter 5

A Robust Method for Detecting Lane Boundary in Challenging Scenes

Trung-Thien Tran, Hyo-Moon Cho and Sang-Bock Cho

Graduate School of Electrical Engineering, Daehaka-ro, Nam-gu, Ulsan 680-749, South Korea

ABSTRACT

Lane boundary detection plays a key role in the driver assistance system. This study proposed a robust method for detecting lane boundary in challenging scenes. First, a horizontal line is detected from the original image using improved Vertical Mean Distribution Method (iVMD) and the sub-region image which is under the horizontal line, is determined. Second, we extracted the lane marking from the sub-region image using Canny edge detector. Finally, K-means clustering algorithm classified left and right lane cluster before using RANSAC algorithm which fits a line model to each cluster. The proposed algorithm demonstrates the accuracy with respect to variant illumination, cracked road, complex lane marking and passing traffic. Experimental results show that the proposed method satisfies the real-time and efficient requirement of the intelligent transportation systems.

1. INTRODUCTION

According to Traffic Accident Analysis Center (http://eng.koroad.or.kr/Eng/Code/RotaStatistic/RotaStatistic10.jsp), there were 231,990 traffic accidents in 2009 across South Korea, killed 5,525 people and injured 226,465 people. The number of accidents rose 2.5% from 2008, with traffic injuries increasing 3.1%. A remarkable number of this statistical information was due to not only the inattention of drivers but also the low road condition. Lane boundary detection helps reduce this huge number of human fatalities. Therefore, lane boundary detection is an important component of many intelligent transportation systems and is used for vehicle navigation, lateral control, collision prevention or lane departure warning system.

Nowadays, many researchers have been researching on lane detection and lane following. Some systems (McCall and Trivedi, 2006; Ma et al., 2000) detected lane boundary by multisensor fusion (RADAR, LASER and Global-positioning System (GPS) but these systems worked with high cost and required updated data frequently in Geographic Information System (GIS) or Route Network Definition File (RNDF). Other researchers applied vision-based systems for lane detection and lane following. However, most of them demonstrated limitations with shadow, variant illumination and artificial symbols on the road surface. For example, several lane detection methods based on top-view image converted from 2D images (Mallot et al., 1991; Aly, 2008). These methods eliminated the perspective effect in the image and focus attention on the sub-region image. Nonetheless, online computation, camera calibration and parallel lane assumption was indispensable and time-consuming process. Recently, some researchers have attempted to implement some methods that are much more robust against noise and missing data (Wang et al., 2004; Jung and Kelber, 2005; Zhou et al., 2006; Cheng et al., 2006; ZuWhan, 2008). These techniques utilized mathematical models to fit data considered as lane boundaries. Nevertheless, they took much computation time and demanded some prerequisite constraints (parallel lane, fixed width and flat ground).

Motivated by the above problems, we proposed a robust method for detecting lane boundary in challenging scenes. The main properties of our method are described as the following.

- This method utilizes a novel method which detects horizontal line from grayscale image. This stage reduces computation time in comparison with processing whole image and is more useful than other methods

(Aly, 2008; Wang et al., 2004; Jung and Kelber, 2005; Zhou et al., 2006; Cheng et al., 2006; ZuWhan, 2008) which used top-view image for same goal. Because the proposed method focuses attention only on a sub-region image and not requires camera calibration, flat ground assumption. Moreover, we apply tracking for horizontal line detection to improve the accuracy of this stage

- Almost accidents often happen in the near field in front of vehicle on highways and urban streets
- Therefore, linear model which is robust to various road conditions, is utilized for expressing lane boundaries. This linear model is simpler and faster than complex model used in (Aly, 2008; Wang et al., 2004; Jung and Kelber, 2005; Zhou et al., 2006; Cheng et al., 2006)
- A combined approach of K-means clustering and Random Sample Consensus (RANSAC) algorithm is applied to determine lane boundaries. This method detects left and right lanes, respectively while some previous approaches (Wang et al., 2004; Jung and Kelber, 2005; Zhou et al., 2006) supposed a fixed-width lane shape

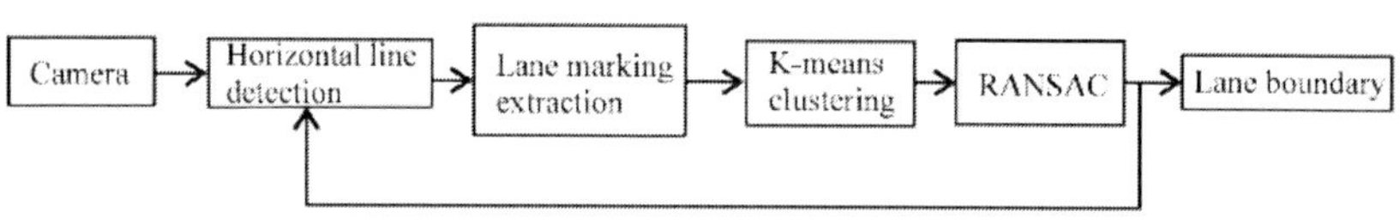

Figure 1. Block diagram of the proposed algorithm

A block diagram of the propose algorithm is shown in Figure 1. Each block of the diagram will be described later.

2. HORIZONTAL LINE DETECTION

A typical camera is installed with the resolution of 240x320 pixels. This stage will determine the sub-region image which is under the horizontal line. This has two benefits. First, the computation time for whole process decrease remarkably because only the sub-region image is considered.

Second, this stage eliminates the bad influence of upper part of image (blue sky in day-time, artificial light in night-time) on lane markings. The previous method (Lim et al., 2009) utilized Vertical Mean Distribution method to determine the first minimum that occurs from the upper curve. That minimum position was considered as the horizontal line position. Vertical Mean Distribution (VMD) is defined as the following equation.

$$VMD(i) = \frac{1}{W}\sum_{j=1}^{W} Gr(i,j); i \in (1,H)$$

where, VMD(i) is the average of the gray values of the row ith. H is the height of original image. W is the width of original image. Gr(i,j) is the intensity of the pixel at the row ith and the column jth in grayscale image converted from original image.

Although, the previous method (Lim et al., 2009) worked well in day-time scenes, it was failed in night-time scenes. In this study, improved Vertical Mean Distribution method (iVMD) is proposed and used for detecting the horizontal line in day-time and night-time scenes.

With day-time scenes, the minimum position, where the first minimum occurs from the upper curve, is considered as the position of horizontal line (red star) shown in Figure 2b. This is because sky region usually possesses higher intensity than road pixel and it may have a big transition of intensity difference as sky approaches ground.

$$\text{Horizontal position} = \arg\min_{i\in(1,H)} VMD(i)$$

With night-time scenes, conversely, the maximum search along the vertical mean curve is considered as the position of horizontal line (green star) shown in Figure 2d. This is because the upper part of the image often lack of illumination and the intensity of horizontal line is highest. The results are shown in Figure 2a-c.

$$\text{Horizontal position} = \arg\min_{i\in(1,H)} VMD(i)$$

However, this stage only applies to first frame of frame sequence or lane boundary detection failed. For next frames, the position of the intersection between two lane boundaries can be considered as a horizontal line position. This is the time-saving and fiducial property of our proposed algorithm.

3. LANE BOUNDARY DETECTION ALGORITHM

Lane marking extraction: In this stage, we ascertain where the lane markings are from the grayscale sub-region image which is cropped from horizontal line position to the bottom of original image and is shown in Figure 3a. The intensity of lane markings is usually higher than the ones of the road surface. Based on this characteristic, Canny edge detector (Canny, 1986) is applied to detect edge points in the grayscale sub-region image.

In complex road, it is difficult to choose a suitable threshold value that eliminates edge points without removing many edge points of lane markings. A low threshold value can be insured the existence of true edge points corresponding to the lane marking or lane boundaries. Therefore, we extracted the edge points with a low threshold value of Canny edge detector. These edge points include lane marking candidates and noise factors of the complex road scene shown in Figure 3b.

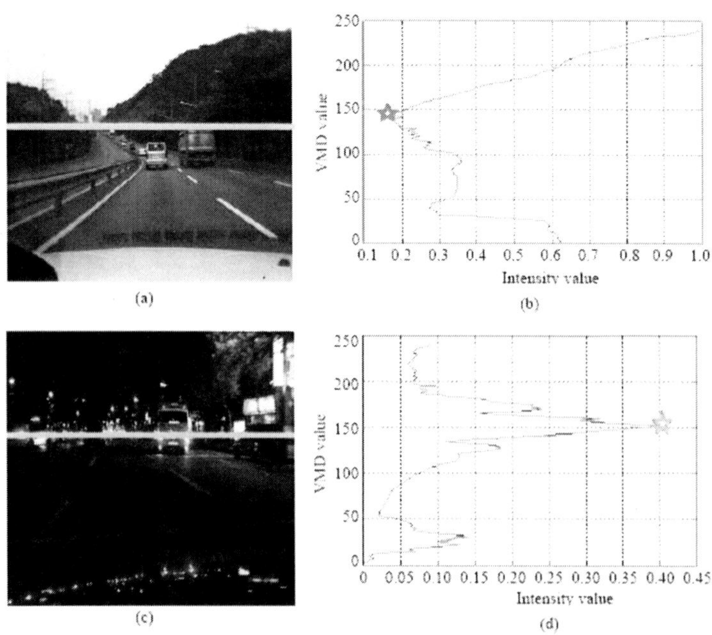

Figure 2. (a-d). Horizontal line detection algorithm

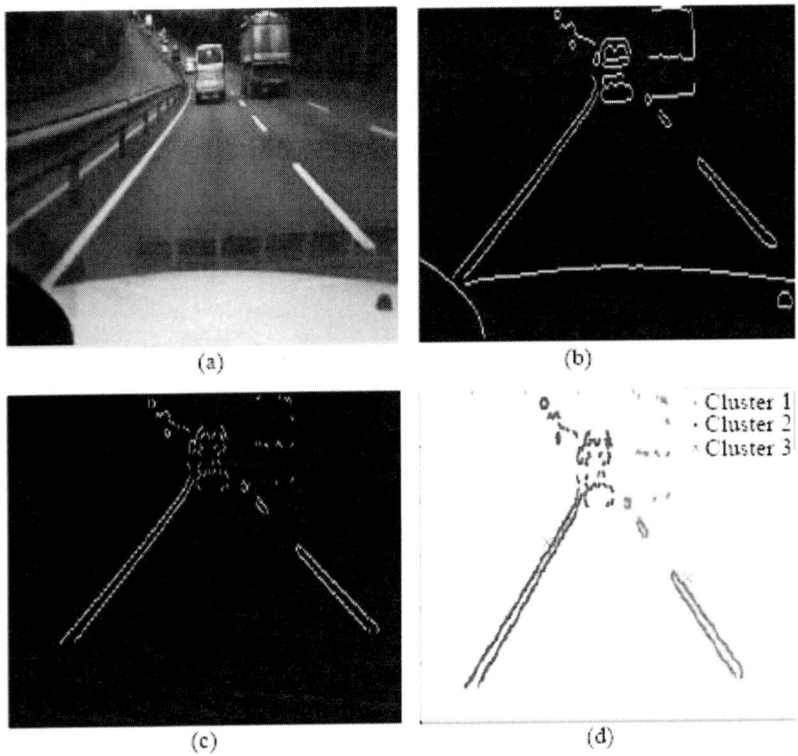

Figure 3. (a-d). (a) Grayscale sub-region image, (b) Canny edge Image, (c) Improved edge image. (d) K-means clustering

The lane marking contains the edge points which are continuity, so we utilized this feature to eliminate some wrong edge points made by noise factors shown in Figure 3c.

K-means and RANSAC algorithm for detecting lane boundary: The proposed method detects left lane and right lane independently. That an edge point belongs to whether left lane or right lane is unknown, so K-means clustering algorithm is used to classify the edge points into two groups: left lane and right lane.

K-means clustering (Kanungo et al., 2002) is a very simple and effective clustering algorithm which requires the number of sought clusters before classifying. K-means clustering aims at minimizing an objective function which is defined as the following.

$$J = \sum_{j=1}^{k} \sum_{i=1}^{n} \| x_i^{(j)} - c_j \|^2$$

where, ||xi(j)-cj ||2 is a chosen distance measure between a data point xi(j) and the cluster centre cj is an indicator of the distance of the n data points from their respective cluster centres. Figure 4 condenses the main stages of K-means clustering algorithm.

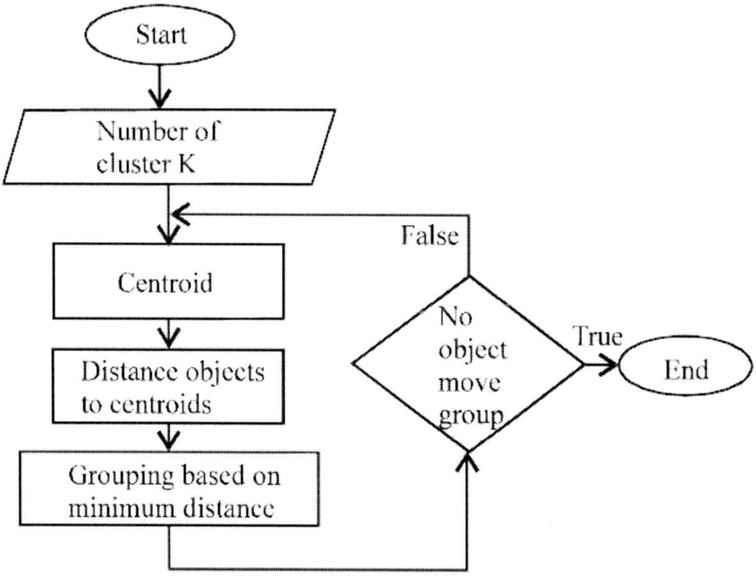

Figure 4. K-means Clustering Algorithm

In present study, we choose k = 2 as the number of clusters. First, 2 edge points are selected randomly as initial cluster centers of each lane. The other edge points are assigned to their closest cluster center according to the distance measured by distance, orientation. In the following step, the means of each cluster are calculated and these means are taken as the new cluster center. Finally, the whole process is repeated with new cluster centers until no change in cluster center. Consequently, we get the left lane set and right lane set, as shown in Figure 3d.

How we can fit a line model to each K-means clustering data set which contains so many outliers shown in Figure 3d. Both Hough Transform and Least square fitting are failed in this situation. Therefore, we use Random Sample Consensus Algorithm (Fischler and Bolles, 1987; Lopez et al., 2005) (RANSAC) which is often used in computer vision when there are a lot of outliers, to determine the parameters of line model. The RANSAC algorithm is briefly described in Figure 5.

The number of iterations, N, is chosen high enough to ensure that the probability (set to 0.99) that at least one of the sets of random samples does not include an outlier. In this study, N = 300 and the final result of proposed algorithm is shown in Figure 6.

1. Selects randomly the minimum number of points required to determine the model parameters.
2. Solve for the parameters of the model
3. Determine how many points from the set of all points fit with a predefined
tolerance E
4. If the fraction of the number of inliers over the total number points in the set exceeds predefined threshold T, re-estimate the model parameters using all the inliers and terminate.
5. Otherwise, repeat steps 1 thorough 4 (maximum of N times)

Figure 5. RANSAC Algorithm

After left and right lane boundaries are detected, the intersection point between them is determined. And then this point is considered as a horizontal line position for next frame of sequence. This work helps to improve the robustness of whole process and reduce computation time because of skipping horizontal line detection stage.

Figure 6. The final result of proposed method

4. EXPERIMENTAL RESULTS

This section shows some experimental results of the proposed algorithm. We applied this proposed algorithm for several single frames and frame sequences to test the performance. Some limitations are exhibited in certain situation.

Single frame results: We collected some images in challenging scenes: cracked road, night scene, rainy road with solid lane or dashed lane. The three first road images (Figure 7a-c) contain cracked road and complex lane marking. The bad qualification of lane marking makes lane detection difficult. The next two images (Figure 7d, e) describe the overtaking of another vehicle or the following of another vehicle. Other vehicles sometimes occlude a part of lane marking or the colors of other vehicles are same as lane marking. The four images (Figure 7f, g, h, i) present some bridges in upper part of the image, or a vehicle run on night scenes.

The lack of light in highway makes the lane detection impossible; the artificial lamp, traffic light and tail-light are bad influence to lane marking.

The three final images (Figure 7j, k, l) is recorded in rainy, fog scenes. The watermark on the windshield or on road surface is the factor which makes lane marking wrong. Nevertheless, the proposed algorithm still gets accurate results in difficult scenes.

Figure 7 (a-l). Some results of proposed method in challenging scenes

Frame sequence results: Above part, we only tested the proposed algorithm with single frame. To demonstrate the effect of the proposed algorithm, we test it on some frame sequences. Therefore, we recorded a number of clips on highway and Metropolitan Ulsan city in day-time and night-time. The system was prototyped using MATLAB R2010a with the Intel Core Quad processor at 2.67 GHz 2GB RAM. The resolution of these clips is (240x320) pixels and the frame rate is 30 FPS.

Clip .1 is collected under day-time condition on highway. This clip consists of different types and colors of lane marking: solid or dashed, yellow or white. Moreover, it contains some difficult scenarios: including overtaking (frame 307), incoming vehicle (frame 964, 1099) and lane changing (frame 380, 421). The detection results are displayed in Figure 8. We observe that the lane boundaries are extracted successfully, regardless of complex scenes.

Clip .2 is captured under day-time on Metropolitan Ulsan city. Almost lane markings are dashed line and white color but there are many complex symbols and writing characters (frame 28,420,443). Sometimes, the lane boundaries are affected by other vehicles (bus, frame 170; incoming vehicle in other lanes, frame 317) and the traffic light, building block. What Figure 9 exhibit the boundary lines of the lane are all correctly determined.

The proposed algorithm also is tested with video clip recorded at night time. The lane boundary detection is not only fatigued task but also more and more difficult during nighttime. Almost lane markings are affected by symbols, tail light and artificial light. Again, Figure 10 shows the precise results of the algorithm.

Some previous methods (Aly, 2008; Wang et al., 2004; Jung and Kelber, 2005; Zhou et al., 2006; Cheng et al., 2006; ZuWhan, 2008) only worked under 10 frames/s speed but our proposed algorithm works with high speed at least 27 frame/s and can correspond to real-time requirement. Moreover, this method demonstrates its high performance over 90%, as shown in Table 1.

Limitations: In the current algorithm, we only focus on current lane which the vehicle is running on and straight line. Although, this algorithm gets surpassing results in difficult scenes, it still has some limitations.

Figure 8. The results from video clip .1

Figure 9. The results from video clip .2

Figure 10. The results from video clip .3

Cases like there are so many symbols on road (frame 1405), the lane marking does not exist (frame 313) and the car encounters the crossing (frame 654), the proposed algorithm may give erroneous results shown in Figure 11.

Figure 11. Some erroneous results of each video clip

Table 1. Results of proposed algorithm

Clip	Total frame	Detected frame	Correct rate (%)	Frame/sec
1	1500	1360	90.66	27.0
2	486	458	94.24	27.0
3	885	756	85.42	28.0
Total	2871	2574	90.11	27.3

5. CONCLUSION

In this study, a robust and real-time method for lane boundary detection is proposed. This method takes horizontal line using improved Vertical Mean Distribution to reduce computation time and to improve the method accuracy. The lane markings are extracted by Canny edge detector with low threshold value which guarantee that all lane markings are detected. After that, K-means clustering and RANSAC fitting technique are utilized to detect lane boundaries. The left and right lane is ascertained separately without some assumptions (parallel and fixed-width lane) and pre-settings (camera calibration). Our proposed algorithm not only gets accurate results but also can correspond to real-time requirement of the intelligent transportation systems.

Further work will extend other models to describe arbitrary shape of the lane boundary. Lane tracking, estimation orientation of the vehicle and lane departure warning system will be implemented. The system would become more robust and safe.

REFERENCES

1. Aly, M., 2008. Real time detection of lane markers in urban streets. Proceeding of the IEEE Symposium on Intelligent Vehicles, June 4-6, Eindhoven, pp: 7-12.

2. Canny, J., 1986. A computational approach to edge detection. IEEE Trans. Pattern Anal. Mach. Intell., 8: 679-698.

3. Cheng, H.Y., B.S. Jeng, P.T. Tseng and K.C. Fan, 2006. Lane detection with moving vehicles in the traffic scenes. Intell. Transport. Syst., 7: 571-582.

4. Fischler, M.A. and R.C. Bolles, 1987. Random Sample Consensus: A Paradigm for Model Fitting with Applications to Image Analysis and Automated Cartography. In: Readings in Computer Vision: Issues, Problems, Principles, and Paradigms, Martin, A.F. and F. Oscar (Eds.). Morgan Kaufmann Publishers Inc., London, pp: 726-740.

5. Jung, C.R. and C.R. Kelber, 2005. Lane following and lane departure using a linear-parabolic model. Image Vision Comput., 23: 1192-1202.

6. Kanungo, T., D.M. Mount, N.S. Netanyahu, C.D. Piatko, R.S. Angela and Y. Wu, 2002. An efficient k-means clustering algorithm: Analysis and implementation. IEEE Trans. Pattern Anal. Mach. Intell., 24: 881-892.

7. Kim, Z., 2008. Robust lane detection and tracking in challenging scenarios. IEEE Trans, Intell. Transp. Syst., 9: 16-26.

8. Lim, K.H., K.P. Seng, L.M. Ang and S.W. Chin, 2009. Lane detection and Kalman-based Linear-parabolic lane tracking. Proceeding of the International Conference on Intelligent Human-Machine Systems and Cybernetics, Aug. 26-27, Hangzhou, Zhejiang, pp: 351-354.

9. Lopez, A., C. Canero, J. Serrat, J. Saludes, F. Lumbreras and T. Graf, 2005. Detection of lane markings based on ridgeness and RANSAC.

Proceedings of the 2005 IEEE International Conference on Intelligent Transportation Systems, Sept. 13-15, Vienna, Austria, pp: 254-259.

10. Ma, B., S. Lakshmanan and A.O. Hero, III, 2000. Simultaneous detection of lane and pavement boundaries using model-based multisensor fusion. Intell. Transport. Syst., IEEE Transact., 1: 135-147.

11. Mallot, H.A., H.H. Bulthoff, J.J. Little and S. Bohrer, 1991. Inverse perspective mapping simplifies optical flow computation and obstacle detection. Biol. Cybernetics, 64: 177-185.

12. McCall, J.C. and M.M. Trivedi, 2006. Video-based lane estimation and tracking for driver assistance: Survey, system, and evaluation. Intell. Transport. Syst. IEEE Transact., 7: 20-37.

13. Wang, Y., E.K. Teoh and D.G. Shen, 2004. Lane detection and tracking using B-Snake. Image Vision Comput., 22: 269-280.

14. Zhou, Y., R. Xu, X. Hu and Q. Ye, 2006. A robust lane detection and tracking method based on computer vision. Measur. Sci. Technol., 17: 736-745.

Chapter 6

Foreground Object Detection by Motion-based Grouping of Object Parts

Zhipeng Wang [1], Jinshi Cui[2], Hongbin Zha[2], Masataka Kagesawa[1], Shintaro Ono[1], Katsushi Ikeuchi[1]

[1]Institute of Industrial Science, The University of Tokyo, Tokyo, Japan

[2]Key Lab of Machine Perception (Ministry of Education), Peking University, Peking, China

ABSTRACT

Effective video-based detection methods are of great importance to intelligent transportation systems (ITS), and here we propose a method to localize and label objects. The method is able to detect pedestrians and bicycle riders in a complex scene. Our method is inspired by the common fate principle, which is a mechanism of visual perception in human beings, and which states tokens moving or functioning in a similar manner tend to be perceived as one unit. Our method embeds the principle in an Implicit Shape Model (ISM). In our method, keypoint-based object parts are firstly detected and then grouped by their motion patterns. Based on the grouping results, when the object parts vote for object centers and labels, each vote belonging to the same object part is assigned a weight according to its consistency with the votes of other object parts in the same motion group. Afterwards, the peaks, which correspond to detection hypotheses on the Hough image formed by summing up all weighted votes, become easier to find. Thus our method performs better in both position and label estimations. Experiments show the effectiveness of our method in terms of detection accuracy.

KEYWORDS

Object detection, Motion grouping, Common fate

1 INTRODUCTION

In ITS areas, detection methods using cameras can be used for navigation, safe driving, surveillance, and sustaining results from other sensors. In traditional ITS applications, vehicles are the main targets. Currently pedestrians are also considered as important subjects of ITS applications, and bicycles also are becoming very popular for environmental and economical reasons. In Japan, the number of traffic accidents among bicycles and pedestrians is very large. Thus we tackle an issue of detecting freely moving bicycle riders and pedestrians from the data collected by a camera which keeps them under surveillance from the top. These situations can be observed in parks, university campuses, station squares, tourist spots, etc. Here we focus on techniques from the area of computer vision for detection under surveillance scenarios. It is also assumed by the method that target objects captured by the surveillance cameras do not change much in scale.

Most state-of-the-art visual detection methods fall into two main categories: sliding-window methods and Hough transform based methods. The methods [10, 27] based on a sliding window schema perform detection in a typical machinery way. In these methods, decisions of whether a target object exists or not are made for part of or all of the sub-images in a test image. Besides the attractive performance and the extendibility of combining various kernels, these methods are favorable because they consider each object as a whole during detection. However, they share limited aspects with visual perception in human beings, and their efficiency heavily relies on the size of the test images.

The other methods [13, 5, 6, 18] detect objects based on the generalized Hough transform [1]. Object parts are detected, and the object parts provide confidence of the locations being the potential objects' centers. Locations of objects are decided according to the converged confidence. These methods are favorable for their robustness to partial deformation and ease of training. To human beings, this kind of method seems to be more natural. And in our work, we combine a mechanism of visual perception in humans, with the ISM [13] to demonstrate this natural property.

A typical Hough transform based method contains two steps: training and detection. During training, a codebook of object parts is built from a set of well annotated images. Each code in the codebook contains information about the appearance of the object part, the relative position to the object center, and the class label. Each object part's appearance is given in the form of keypoint descriptors [13], image patches [7, 19], or image regions [8]. Each code not only encodes one object part's appearance, but also its offset to the object center and the class label. During the detection step, object parts are detected on each test image. Then every object part is matched against the codebook, and several codes nearest in appearance are activated. The offset and class label encoded in each activated code will act as a vote. All the votes from the object parts are added up to form a Hough image. The peaks of the Hough image are considered detection hypotheses with the height of each peak as the confidence for the corresponding hypothesis.

Two challenging issues for detection methods are how to separate near objects and how to separate similar different-class objects. The target objects, in the case of ITS applications, are pedestrians, bicycle riders, and automobiles. In the schema of sliding window, usually non-maximum suppression is needed for post-processing, and a mechanism in [10] works by excluding from the feature pool the features which belong to each successive detection response. In Hough transform based methods, a similar mechanism is also employed in [2], however, this effort is employed after the forming of a Hough image. During the forming of a Hough image, two kinds of votes make detection challenging: (1) votes cast by object parts from near objects make the peaks corresponding to different objects mixed up, and (2) votes cast by similar different-class object parts lead to tough decisions on the class label of the peaks. See Figure 2d. Before the forming of Hough images, problems also arise from the pollution of the training images' background part to the codebook. During training a very clean codebook can be built with the foreground marked, which requires manual efforts. Otherwise, a large amount of training examples are needed for the effectiveness of the codebook, and this decreases efficiency.

In videos, motion information is also available by simple tracking of object parts. Thus we propose a method for detection which utilizes both appearance and motion information. The method is based on the common fate principle [23]. The principle is one of the visual perception principles as theorized by gestalt psychologists, and it states that for human beings, tokens moving coherently are perceptually grouped. This provides an intuition to group the object parts by their motion patterns, and let them vote afterwards. In our work, the object parts are represented using keypoint descriptors, which are tracked to generate trajectories. The object parts are grouped by the pairwise similarities of their corresponding trajectories. Using the assumption that object parts in the same motion group probably belong to the same object, for each object part, we assign higher weights for the votes of the object parts which are more "agreeable" within the motion group. This results in votes corresponding to true detection responses to be more likely assigned higher weights. And on a Hough image formed by summing up these weighted votes, the peaks are easier to find as shown in Figure 1d.

Due to the combination of motion analysis results and the Hough transform framework, and by assigning different weights to each object part's votes, the proposed method has several appealing properties:

- The method's ability to estimate object position and label multiple objects from different classes. The existence of three types of objects makes the task challenging: near objects, similar different-class objects, and multi-pose same-class objects.
- Its ability to use a codebook trained by images with cluttered backgrounds.
- The framework used to combine grouping results of object parts is very general, and thus can be easily expanded.

The remaining paper is organized as follows. Section 2 reviews related work. Section 3 formalizes the common fate Hough transform. Section 4 describes inference on the formed Hough images. Section 5 gives experimental results, and Section 6concludes.

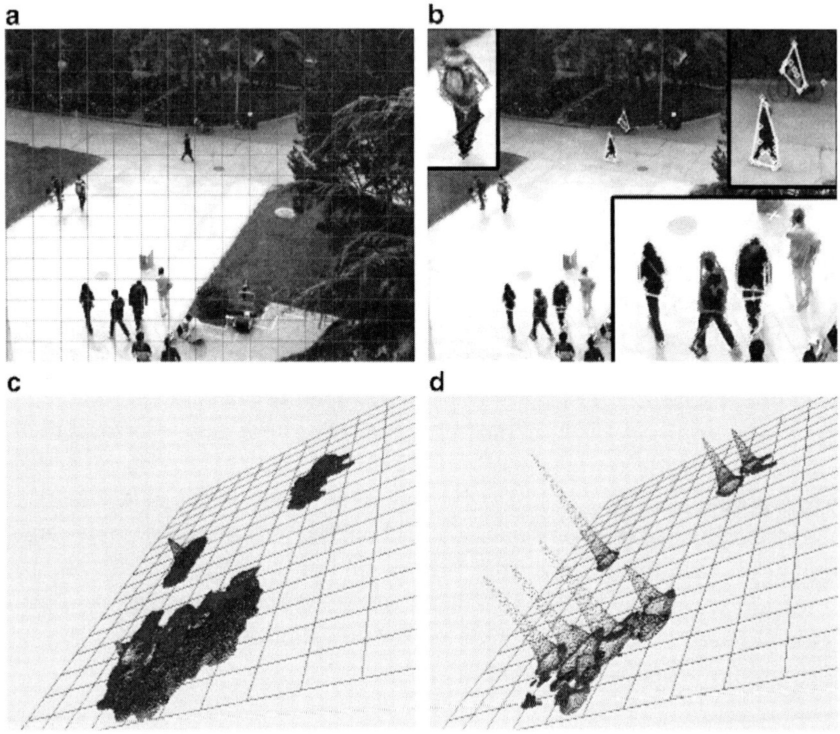

Figure 1. Merit of the proposed method. **a** Original image. **b** Motion grouping results. Some parts are enlarged to show details. **c** Original Hough image. **d** Hough image formed using our method. The grids in **c** and **d** correspond to the grids in **a**

2 RELATED WORK

Our work is most closely related to object detection methods [2, 12, 13, 14, 15, 16] based on the Hough transform framework. Recently, such methods are making a lot of progress. The ISM [13, 14] is extended by notifying correspondences between the object parts and the hypotheses [2] for the detection of multiple near objects. While in the methods [7, 15, 19], the Hough transform is placed in a discriminative framework for object detection in a way that the codes are assigned different weights by the co-occurrence frequency of their appearance and offset to the object center. Two Hough transform methods consider the grouping of object parts [20, 26]. The method in [20] deals with scale change. Instead of estimating the scale by local features trained from different scaled examples, the votes are considered as voting lines. By considering the difference between the voted centers, local features are first grouped, resulting in a more consistent vote for the object center. In [26], the grouping of object parts, the correspondence between object parts and object, and the decisions on detection hypotheses are optimized in the same energy function. For this method, the problem is that the grouping results don't have meaning or correspond to any real entities.

Our work is also related to object detection methods which use trajectories [3, 4], methods using the weighting of features [25], methods dealing with codebook noise [17], and methods which integrate temporal information [24].

3 COMMON FATE HOUGH TRANSFORM

Probabilistic standpoints are very appealing because of inference ease. However, as pointed out in [11], placing an Implicit Shape Model (ISM) in a probabilistic framework is not satisfactory. Especially, describing weights of the votes as priors does not make sense. A Hough transform can be simply considered as the transformation from a set of object parts, $\{\mathbf{e}\}$, to a confidence space of object hypotheses, $C(\mathbf{x}, l)$. Where $\mathbf{x}$ is the coordinate of the object center, and l the label. Terms described as priors of the votes in the ISM are actually weights, and the likelihood terms are actually blurring functions to convert discrete votes into continuous space. This section describes how a Hough image for the estimation of object centers and labels is formed from object parts observed on an image.

Let $\mathbf{e}$ denote an object part observed on the current image. The appearance of $\mathbf{e}$ is matched against the codebook, and $\mathbf{e}$ activates N best matched codes from the trained codebook. Each code contains the appearance, its offset to the object center, and the class label. According to the N matched codes, $\mathbf{e}$ casts N votes. Each vote $V_{\mathbf{e}}$ is about the object center that generates $\mathbf{e}$. The position of the object center casted by a vote, V, is denoted by $\mathbf{x}_V$, while the class label is l_V. Based on the N votes of $\mathbf{e}$, the confidence that a position x~x~ is the center of an object with class label l~l~ is given by,

$$C(\tilde{\mathbf{x}}, \tilde{l}; \mathbf{e}) = \sum_{i=1}^{N} B\left(\tilde{\mathbf{x}}, \tilde{l}; V_{\mathbf{e}}^{i}\right) w\left(V_{\mathbf{e}}^{i}\right). \qquad (1)$$

Here B $(\tilde{\mathbf{x}}, \tilde{l}; V_{\mathbf{e}}^{i})$ is the blurring function. And w (Vie) is the weight of V_e^i.

The idea of the proposed method is that the weight term, $w(V^i_e)$, is defined by the motion grouping results of all the object parts.

The blurring function is defined as,

$$B(\tilde{\mathbf{x}}, \tilde{l}; V) = \begin{cases} 0 & \text{if } l_V \neq \tilde{l} \text{ or } |\tilde{\mathbf{x}} - \mathbf{x}_V| > d \\ G(\tilde{\mathbf{x}}; \mathbf{x}_V, \sigma) & \text{otherwise} \end{cases} \tag{2}$$

Here $G(\tilde{\mathbf{x}}; \mathbf{x}_V, \sigma)$ is a Gaussian function that fixes the spatial gap between x~ and $\mathbf{x}_V$.

Let M be the total number of object parts on the image, then by summing up over all the object parts, the confidence of x~ being the center of an l~-class object is given by,

$$\begin{aligned} C(\tilde{\mathbf{x}}, \tilde{l}) &= \sum_{j=1}^{M} C(\tilde{\mathbf{x}}, \tilde{l}; \mathbf{e}_j) w(\mathbf{e}_j) \\ &= \sum_{j=1}^{M} \sum_{i=1}^{N} B\left(\tilde{\mathbf{x}}, \tilde{l}; V_{\mathbf{e}_j}^{i}\right) w\left(V_{\mathbf{e}_j}^{i}\right) w(\mathbf{e}_j) . \end{aligned} \tag{3}$$

A uniform weight is assumed for each object part, and $w(\mathbf{e}_j) = \frac{1}{M}$. By considering $C(\tilde{\mathbf{x}}, \tilde{l})$ as the evaluation score of the Hough space $(\tilde{\mathbf{x}}, \tilde{l})$, the task of estimating object centers and labels converts to finding, and then validating, the local maxima of the Hough image.

3.1 Common Fate Weights

To meet the challenges of separating near objects, separating similar different-class objects, and using a noisy codebook, different weights are assigned to the votes of each object part by considering the motion grouping results of the object parts. In this sub-section, when given some grouping results, how the results are combined into a Hough transform framework is introduced.

Let γ = {**g**} denote the grouping results, where **g** is a group of object parts. Assume $\mathbf{e}_m \in \mathbf{g}$ and $\mathbf{e}_n \in \mathbf{g}$. Those votes of $\mathbf{e}_m$ which are more "agreeable" than the votes of the other objects in **g** are assigned larger weights.

Towards this end, the relationship between the votes of $\mathbf{e}_m$ and the votes of $\mathbf{e}_n$ needs to be given in advance. This relationship is named support. The support from VenVen to VemVem is defined based on VenVen and the confidence that VemVem's voted center is correct, as,

$$S(V_{\mathbf{e}_n} \rightarrow V_{\mathbf{e}_m}) = B(\mathbf{x}_{V_{\mathbf{e}m}}, l_{V_{\mathbf{e}m}}; V_{\mathbf{e}_n}) \, , n \neq m \, .$$

Here $B(\mathbf{x}_{V_{\mathbf{e}m}}, l_{V_{\mathbf{e}m}}; V_{\mathbf{e}_n})$ is defined in Eq. 2. This measures the coherence of the two votes from different object parts.

Then, the support from $\mathbf{e}_n$ to VemVem is defined based on $\mathbf{e}_n$, and the confidence that VemVem's voted center is correct, as,

$$\begin{aligned} S(\mathbf{e}_n \rightarrow V_{\mathbf{e}_m}) &= C(\mathbf{x}_{V_{\mathbf{e}m}}, l_{V_{\mathbf{e}m}}; \mathbf{e}_n) \\ &= \sum_{i=1}^{N} S\left(V_{\mathbf{e}_n}^i \rightarrow V_{\mathbf{e}_m}\right) w\left(V_{\mathbf{e}_n}^i\right) , n \neq m \, . \end{aligned}$$

And the support from **g** to VemVem is defined by the confidence that VemVem's voted center is correct based on the votes of all the other object parts excluding its belonging object part in **g**, as,

$$\begin{aligned} S(\mathbf{g} \rightarrow V_{\mathbf{e}_m}) &= \sum_{\mathbf{e}_i \in \mathbf{g} - \{\mathbf{e}_m\}} C(\mathbf{x}_{V_{\mathbf{e}_i}}, l_{V_{\mathbf{e}m}}; \mathbf{e}_i) w(\mathbf{e}_i) \\ &= \frac{1}{M} \sum_{\mathbf{e}_i \in \mathbf{g} - \{\mathbf{e}_m\}} S(\mathbf{e}_i \rightarrow V_{\mathbf{e}_m}) \, . \end{aligned}$$

By assuming all object parts in the same motion group are from the same object, which means motion grouping gives good results, the estimations for center position and class label given by every object part should be consistent with that given by the motion group. Thus for a particular vote of $\mathbf{e}_m$, i.e., V~emV~em, a weight is assigned to it by considering its consistence with **g** and the consistence of $\mathbf{e}_m$'s other votes with **g**, as

$$w\left(\tilde{V}_{\mathbf{e}_m}\right) = \frac{S\left(\mathbf{g} \to \tilde{V}_{\mathbf{e}_m}\right) + \frac{\Delta}{N}}{\sum_{i=1}^{N} S\left(\mathbf{g} \to V_{\mathbf{e}_m}^{i}\right) + \Delta}$$

$$= \frac{\sum_{\mathbf{e}_j \in \mathbf{g}-\{\mathbf{e}_m\}} \sum_{k=1}^{N} S\left(V_{\mathbf{e}_j}^{k} \to \tilde{V}_{\mathbf{e}_m}\right) w\left(V_{\mathbf{e}_j}^{k}\right) + \frac{M\Delta}{N}}{\sum_{i=1}^{N} \sum_{\mathbf{e}_j \in \mathbf{g}-\{\mathbf{e}_m\}} \sum_{k=1}^{N} S\left(V_{\mathbf{e}_j}^{k} \to V_{\mathbf{e}_m}^{i}\right) w(V_{\mathbf{e}_j}^{k}) + M\Delta}. \quad (4)$$

Here, Δ is a small constant for preventing zeros. Notice $w\left(\tilde{V}_{\mathbf{e}_m}\right)$ is defined using $w\left(V_{\mathbf{e}_j}^{k}\right)$ - the weights of the votes of the other object parts in *g*. In order to determine $w(\tilde{V}_{\mathbf{e}_m})$, uniform weights are firstly assigned to the votes of each object part in **g**, i.e., $w\left(V_{\mathbf{e}_j}^{k}\right) = \frac{1}{N}$. Then new weights are calculated based on the uniformly assigned weights. The weights of votes used to form the Hough image are the iteratively converged weights.

The grouping result $y = \{\mathbf{g}\}$, can be replaced by grouping results based on other information, for example our method utilizes motion to group the voting elements. The manner of extending the Hough transform is very general, and the extended Hough transform with motion grouping results is called the common fate Hough transform. The votes given by the best matched codes and the votes with higher defined weights are shown in Figure 2.

3.2 Motion Grouping

In this subsection, how to group the object parts by their motion patterns is introduced. Basically the object parts are tracked, and clustered by their motion patterns. The object parts are tracked through frames before and after the current frame, to generate trajectories. Then the object parts are grouped by their corresponding trajectories' pairwise motion similarities.

The object parts in this method are in the form of keypoint descriptors. The Harris Corner [9] feature is chosen, for robustness, to represent each object part, while for appearance, the region covariance [22] feature of the image patch around each keypoint is used. The image feature is chosen because of its flexibility to combine multiple channels of information, and also for its capability of handling scale changes in a certain range.

Figure 2. Effect of the proposed weight. **a** Motion groups, different colors mark different motion groups. **b** Voted centers given by the 7 best matched codes. **c** Voted centers with the highest defined weights. **d** Voted centers with weights higher than a threshold

For each object part, a trajectory is generated by tracking its corresponding Harris Corner by the KLT tracker [21]. To group the trajectories, two pairwise similarities are defined.

Let TemTem and TenTen denote two trajectories corresponding to $\mathbf{e}_m$ and $\mathbf{e}_n$. The first similarity between two trajectories is defined as,

$$D_1\left(T_{\mathbf{e}_m}, T_{\mathbf{e}_n}\right) = \max_{i=1\ldots L}\left(\left|\mathbf{x}^i_{T_{\mathbf{e}_m}} - \mathbf{x}^i_{T_{\mathbf{e}_n}}\right|\right).$$

Here, i is the frame index, and L is the number of frames in which both trajectories exist.

To define the second similarity, the *i*th directional vector of *T* is firstly defined as, $\mathbf{d}_T^i = \mathbf{x}_T^{i+3} - \mathbf{x}_T^i$. Let $\mathbf{a}_i = \mathbf{d}_{T_{e_m}}^i$, $\mathbf{b}_i = \mathbf{d}_{T_{e_n}}^i$, $a_i = \frac{\mathbf{a}_i \cdot \mathbf{b}_i}{\mathbf{a}_i \cdot \mathbf{a}_i}$, and $b_i = \frac{\mathbf{a}_i \cdot \mathbf{b}_i}{\mathbf{b}_i \cdot \mathbf{b}_i}$. Then the second similarity is defined as,

$$D_2(T_{\mathbf{e}_m}, T_{\mathbf{e}_n}) = \max_{i=1...L-3} (\max(|\mathbf{a}_i - a_i \mathbf{a}_i|, |\mathbf{b}_i - b_i \mathbf{b}_i|)) .$$

Before grouping the trajectories, the static points are excluded. The defined D_1 is calculated for all pairs of trajectories, and a minimum spanning tree is then built using the calculated similarities. The built minimum spanning tree is split by cutting edges larger than a threshold, D1thDth1, and this gives a grouping result of the trajectories. For each element in the clustering result, D_2 is used in the same procedure to generate even smaller clusters. This hierarchical procedure ensures that trajectories in the same group have both small D_1 and D_2. Max operation is used in the definitions of both D_1 and D_2. This is helpful because very often two trajectories are of different lengths, and under such situations, max operation will have better stability than other operations, e.g., average, that consider only overlapping frames.

Each trajectory corresponds to an object part, and the grouping results of the trajectories correspond to grouping results of the object parts.

3.3 Codebook

For training, Harris corners are extracted from the training images with the object center and the class label annotated. In this method, region covariance is chosen to represent the appearance, which is defined as,

$$\mathbf{r} = \frac{1}{K-1} \sum_{i=1}^{K} (\mathbf{z}_i - \mu)(\mathbf{z}_i - \mu)^T .$$

Here, *K* is the number of pixels in the region, and $\mathbf{z}_i$ is a 7-dimensional vector regarding the (*x*, *y*) coordinate of the pixel, while μ is the mean of $\mathbf{z}_i$. And $\mathbf{z}(x, y)$ contains the RGB color of the pixel and the intensity gradients of the pixel, as: $r(x, y)$, $g(x, y)$, $b(x, y)$, $\left|\frac{\partial I(x,y)}{\partial x}\right|$, $\left|\frac{\partial I(x,y)}{\partial y}\right|$, $\left|\frac{\partial^2 I(x,y)}{\partial x^2}\right|$, and $\left|\frac{\partial^2 I(x,y)}{\partial y^2}\right|$.

The appearance similarity between $\mathbf{r}_m$ and $\mathbf{r}_n$ is given by,

$$\rho(\mathbf{r}_m, \mathbf{r}_n) = \sqrt{\sum_{i=1}^{7} \ln^2 \lambda_i} \; .$$

Here, λ_i is the generalized eigenvalue obtained by solving the generalized eigenvalue problem, $\lambda_i \mathbf{r}_m \mathbf{u}_i = \mathbf{r}_n \mathbf{u}_i$, $\mathbf{u}_i \neq \mathbf{0}$, with $\mathbf{u}_i$ the eigenvector.

A square image patch around each keypoint is used to represent the appearance of an object part. Six region co- variances are generated for each image patch by using the pixels of the top-left, the top-right, the bottom-left, the bottom-right, the central portion, and the entire image patch. Then besides the offset and the class label, a code contains six region covariances. All codes from all training images constitute the codebook. When an object part is matched against the codebook, the similarity between the image patch of the object part and a code is defined by the smallest similarity of the corresponding region covariance. This will handle scale changes of a small range, since the six image patches are not of the same scale. And the method's ability of handling scale changes is limited. So it can only be used in surveillance situations where the scales of target objects change in a limited range.

4 DETECTION

After forming the Hough image, the detection hypotheses are validated. Let $\mathbf{h} = \{H\}$ be the points in the Hough space which are evaluated by $C(\mathbf{x}_H, l_H)$ and have $C(\mathbf{x}_H, l_H) > 0$. Inspired by [2], the hypotheses are validated by an optimizing procedure. Let O be the number of the points in $\mathbf{h}$. Let $u_i = 1$ or 0, indicate H_i as a true object center or not. The problem is:

$$\arg\max_{u_i} \prod_{i=1}^{O} C^{u_i}(H_i) \Longleftrightarrow \arg\max_{u_i} \sum_{i=1}^{O} u_i \ln(C(H_i)) \, .$$

Let $v_{ij} = 1$ or 0, indicate $\mathbf{e}_j$ belongs to H_i or not, then

$$
\begin{aligned}
C(H_i) &= \sum_{j=1}^{M} C(\mathbf{x}_{H_i}, l_{H_i}; \mathbf{e}_j) w(\mathbf{e}_j) \\
&= \frac{1}{M} \sum_{j=1}^{M} v_{ij} C(\mathbf{x}_{H_i}, l_{H_i}; \mathbf{e}_j),
\end{aligned}
$$

and by assuming one object part belongs to and only belongs to one hypothesis, the problem is,

$$
\begin{aligned}
&\arg\max_{u_i, v_{ij}} \sum_{i=1}^{O} u_i \ln \left(\sum_{j=1}^{M} v_{ij} C(\mathbf{x}_{H_i}, l_{H_i}; \mathbf{e}_j) \right) \\
&s.t.: \quad u_i = 0 \text{ or } u_i = 1, \forall\, i; \\
&\qquad\quad v_{ij} = 0 \text{ or } v_{ij} = 1, \forall\, i, \forall\, j; \\
&\qquad\quad \sum_{i=1}^{O} v_{ij} = 1, \forall\, j; \\
&\qquad\quad \sum_{j=1}^{M} v_{ij} \le u_i, \forall\, i.
\end{aligned}
$$

Following [2], the optimal result for the problem is given by greedy maximization. As described in Algorithm 1, the largest local maximum of all the local maxima is chosen to be the center of a true object, and then the object parts belonging to the chosen object center are excluded from the object part set. A new Hough image, where new objects are found, is formed using the remaining object parts. And this procedure ends when the object part set is empty, or when the confidence of the chosen object is lower than a given threshold.

5 EXPERIMENTAL RESULTS

In our experiments, improvement of the method is verified in terms of detection accuracy. The method is tested on the P-campus dataset with [2] as benchmark, and then tested on a dataset of several animals.

Algorithm 1 Greedy Maximization

Let ε be the set of object parts, C_{th} be the low confidence threshold to accept detection responses, and $\hat{\mathbf{h}}$ be the local maxima of $\mathbf{h}$

while $\varepsilon \neq \emptyset$ **do**
  Form $\mathbf{h}$ with ε
  Generate $\hat{\mathbf{h}}$ and select $H_i \in \hat{\mathbf{h}}$ with the largest $C(\mathbf{x}_{H_i}, l_{H_i})$
  if $C(\mathbf{x}_{H_i}, l_{H_i}) >= C_{th}$ **then**
    for $\mathbf{e}_j \in \varepsilon$ **do**
      if $\forall H' \in \hat{\mathbf{h}}, C(\mathbf{x}_{H_i}, l_{H_i}|\mathbf{e}_j) >= C(\mathbf{x}_{H'}, l_{H'}|\mathbf{e}_j)$**then**
        $\varepsilon \leftarrow \varepsilon - \{\mathbf{e}_j\}$
      end if
    end for
  else
    $\varepsilon \leftarrow \emptyset$
  end if
end while
return $\{H_i\}$

5.1 Campus-scene Detection

5.1.1 Dataset

The P-campus dataset contains two primary classes of foreground objects: pedestrians and bicycle riders. The frame size is 720 × 576. Among all the 401 continuous frames, 633 different-class ground truth bounding boxes are annotated on 79 frames. In this dataset, pedestrians and bicycle riders have in common the upper human body, and pedestrians appear in front, back, and side views.

5.1.2 Implementation Settings

For training, 52 bicycle riders and 171 pedestrians are randomly selected from the marked ground truths. Harris corners are detected on these randomly selected training images, examples are given in Figure 3a. For appearance, six region covariances are generated for each keypoint using the 9 × 9 image patch around it as shown in Figure 3b. The appearance, the offset to the image (object) center, and the label of the training image are encoded into a code, and the code is inserted into a codebook. The final codebook contains 5502 codes. Testing data is formed by the 79 frames, on which the ground truth bounding boxes are marked. Harris corners are detected, and region covariances are generated in the same manner as for the training images. For each Harris corner on one testing image, the corresponding region covariances are matched against the codebook for the most similar codes. Some of the training examples will appear in the test sequences. The emphasis of this experiment is to verify the proposed framework's ability of combining motion information. Both the proposed method and the benchmark method use the same training and testing images, so the comparison is fair and proves the effectiveness of the proposed method.

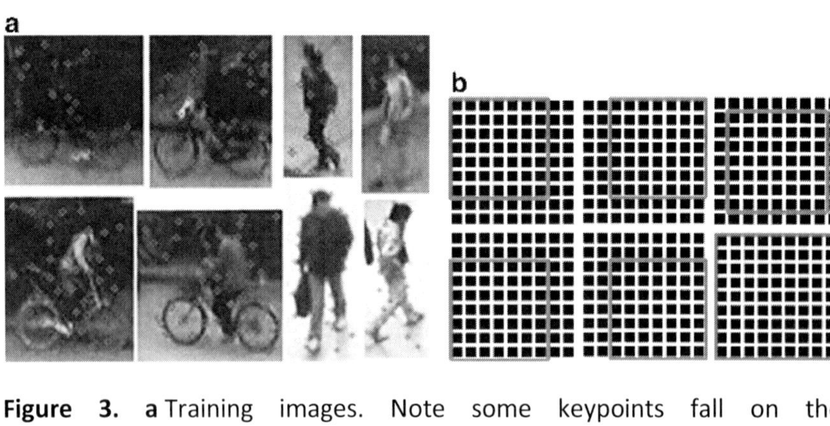

Figure 3. **a** Training images. Note some keypoints fall on the background. **b** The manner how a 9 × 9 image patch is used to generate six region covariances, and *red rectangles* indicate the pixels used for each covariance

For motion grouping, each keypoint is tracked through 10 frames before, and 10 frames after the current frame. The similarity of two 21-point trajectories is defined using only the frames in which both trajectories exist. To set the two thresholds for motion grouping, D_1 and D_2 are measured for keypoint pairs of different objects. D1thDth1is set so that it is larger than only 10% of the measured D_1s, and so is D2thDth2. By doing this, keypoints belonging to different objects are not likely to be grouped together. So that in one motion group, the keypoints are very likely to belong to the same object, as shown in Figure 4.

Figure 4. Motion grouping results

To form the Hough image, 35 best matched codes are chosen from the codebook for each object part. In Eq. 3, d and σ need to be given. The precision-recall curves are based on σ, while d is set to 10. Here σ is the most important parameter.

5.1.3 Comparisons

For comparison, detection is done on the Hough images formed with and without motion grouping results. The same codebook and the same parameter settings are used for forming and searching over both Hough images. The votes of each object part are assigned uniform weights in the benchmark method, while weights defined in Eq. 4 are assigned in the proposed method.

The precision-recall curves are shown in Figure 5a. An object is considered as correctly detected only if the distance from the ground truth to it is less than 10 pixels. In Figure 5a, the correctly positioned but wrongly labeled objects are considered as true positives, aiming at verifying the positioning ability of the proposed method.

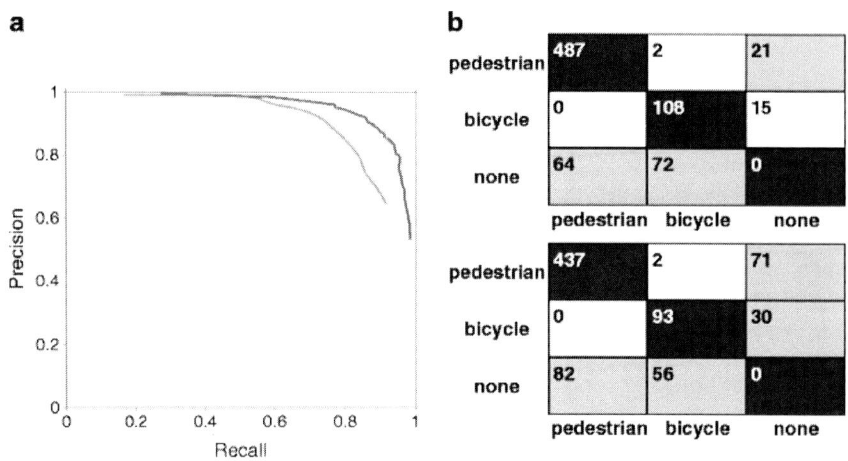

Figure 5. a Precision-recall curves (*red*: the proposed method, *blue*: the benchmark method). **b** Confusion matrices (*upper*: the proposed method, *lower*: the benchmark method)

The confusion matrices are given in Figure 5b. For clarity, the proposed method is compared with the benchmark method when the two methods have a nearly equal number of false alarms. To evaluate the labeling ability, a class of "none" to represent missed detections and false alarms is manually added. For example, in Figure 5b, 487 pedestrian instances are correctly positioned and labeled by the proposed method; 2 are wrongly labeled to be bicycle riders, and 21 are miss-detected. More results are shown in Figure 6.

Figure 6. Results. *Red rectangles* and *blue rectangles* mark correctly detected pedestrians and bicycle riders. *Yellowrectangles* mark missed detections. *White rectangles* mark correctly detected but not correctly labeled objects.*Green rectangles* mark false alarms. *Black rectangles* mark static objects which are beyond the verification of this method

5.2 Wild-scene Detection

5.2.1 Dataset

In order to show that our method can be used for general purposes, we test our method on complicated scenes, especially, complicated background. Even in these cases, our method works well, which shows robustness of our method. A mini dataset is built upon leopards and tigers of the family Felidae. Note especially that the image feature used by this method belongs to the type texture, and texture from different positions of the leopards are almost the same. The dataset contains 6 video clips of 9 leopards and 4 tigers. The frame size is 640 × 480. Both of the animals are in the side view.

5.2.2 Implementation settings

Most implementation settings are the same as the settings used for campus object detection. For training, 5 leopards and 2 tigers are used. The size of the image patch around each keypoint is 27 × 27.

5.2.3 Comparisons

In Figure 7, the motion grouping results, and how the voted centers are affected, are given. Since parts from different positions of the leopard are very similar, the true center of a leopard is difficult to find using the voted centers of the object parts. In Figure 8, example Hough images are given to show the merit of the proposed prior by the ability to detect leopards. In Figure 9, the detection results are given. The proposed method successfully localizes and labels all the leopards and tigers, while the benchmark method miss-detects three leopards.

6 CONCLUSION

The computational ability of human beings is limited, while their ability to detect is far beyond that of machines. Thus, it is very possible that this detection ability benefits from multiple perceptual mechanisms. By using one of these mechanisms, we propose a detection method. By embedding motion grouping results into the voting schema of hough transforms, the method is able to distinguish near objects' positions, distinguish similar objects' labels, and maintain the detection rate with a noisy codebook. The success of our method further demonstrates the advancement of perceptual mechanisms in human beings. And the success of this method will help with detection methods in ITS areas.

ACKNOWLEDGMENTS

This work was, in part, supported by SCOPE program by the Ministry of Internal Affairs and Communications. Part of the work was done during the first author's master course at Peking University. The first author is sponsored by China Scholarship Council. The authors thank Bo Zheng, Boxin Shi, Patricia Knapp, and Nicole Hajicek for suggestions.

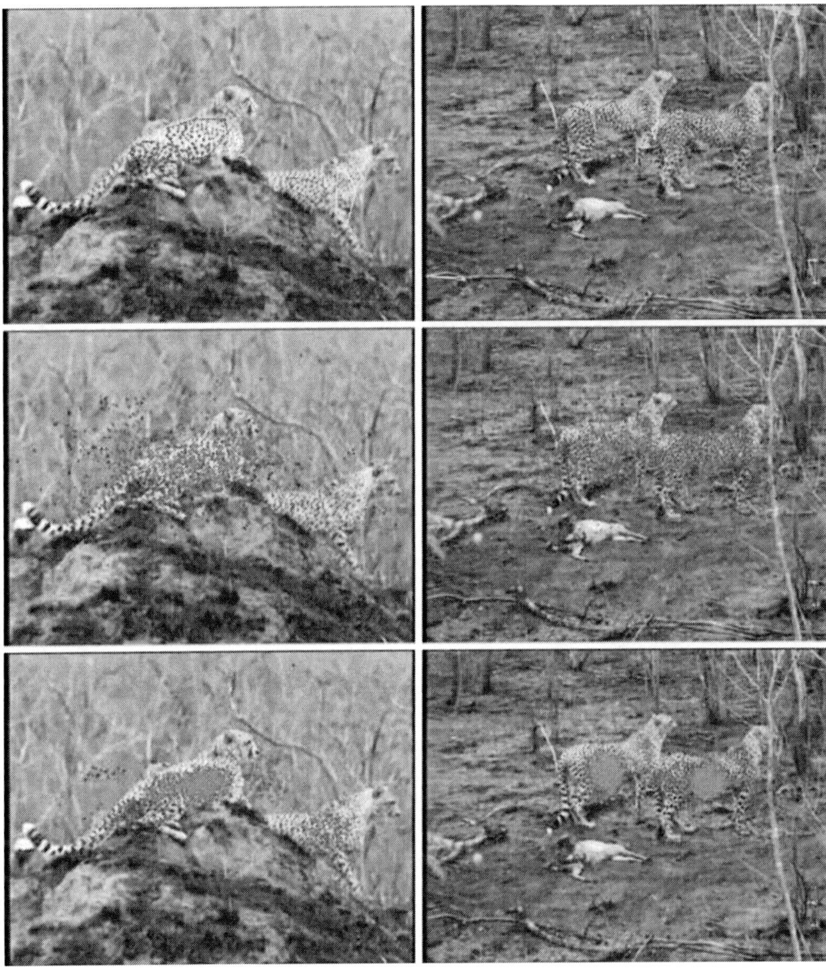

Figure 7. Effect of the proposed weight assignment. *Red circles* are voted centers for leopards, while *blue* ones are voted centers for tigers. On the *top* are the motion grouping results. In the *middle* are the voted centers according to the best matched codes. On the *bottom* are the voted centers voted by votes with the highest weights

Figure 8. Example Hough images. On the *top* are the original images. In the *middle* are the Hough images formed by votes with uniform priors. On the *bottom* are the Hough images formed by votes with the proposed weights.*Red* indicates leopards, and *blue* indicates tigers. Note that for the two leopards, there is no peak corresponding to the one on the *right*, on the benchmark Hough image. For the three leopards, there is also no peak corresponding to the leopard behind on the benchmark Hough image

Figure 9. Results. *Red* crosses mark the centers for leopards and *blue* crosses mark the centers for tigers

REFERENCES

1. Ballard, D.: Generalizing the hough transform to detect arbitrary shapes. Pattern Recog. **13**(2), 111–122 (1981)

2. Barinova, O., Lempitsky, V., Kohli, P.: On detection of multiple object instances using hough transforms. In: CVPR, pp. 2233–2240 (2010)

3. Brostow, G., Cipolla, R.: Unsupervised Bayesian detection of independent motion in crowds. In: CVPR, vol. I, pp. 594–601 (2006)

4. Brox, T., Malik, J.: Object segmentation by long term analysis of point trajectories. In: ECCV, vol. V, pp. 282–295 (2010)

5. Felzenszwalb, P., Huttenlocher, D.: Pictorial structures for object recognition. IJCV **61**(1), 55–79 (2005)

6. Fergus, R., Perona, P., Zisserman, A.: Object class recognition by unsupervised scale-invariant learning. In: CVPR, vol. II, pp. 264–271 (2003)

7. Gall, J., Lempitsky, V.: Class-specific hough forests for object detection. In: CVPR, pp. 1022–1029 (2009)

8. Gu, C., Lim, J., Arbelaez, P., Malik, J.: Recognition using regions. In: CVPR, pp. 1030–1037 (2009)

9. Harris, C., Stephens, M.: A combined corner and edge detector. In: Alvey Vision Conference, pp. 147–152 (1988)

10. Lampert, C., Blaschko, M., Hofmann, T.: Beyond sliding windows: object localization by efficient subwindow search. In: CVPR, pp. 1–8 (2008)

11. Lehmann, A., Leibe, B., Van Gool, L.: Fast prism: branch and bound hough transform for object class detection. IJCV **94**(2), 175–197 (2011)

12. Leibe, B., Cornelis, N., Cornelis, K., Van Gool, L.: Dynamic 3D scene analysis from a moving vehicle. In: CVPR, pp. 1–8 (2007)

13. Leibe, B., Leonardis, A., Schiele, B.: Robust object detection with interleaved categorization and segmentation. IJCV **77**(1–3), 259–289 (2008)

14. Leibe, B., Schiele, B.: Interleaved object categorization and segmentation. In: BMVC, pp. 759–768 (2003)

15. Maji, S., Malik, J.: Object detection using a max-margin hough transform. In: CVPR, pp. 1038–1045 (2009)

16. Mikolajczyk, K., Leibe, B., Schiele, B.: Multiple object class detection with a generative model. In: CVPR, vol. I, pp. 26–36 (2006)

17. Mohottala, S., Ono, S., Kagesawa, M., Ikeuchi, K.: Fusion of a camera and a laser range sensor for vehicle recognition. In: OTCBVS, pp. 16–23 (2009)

18. Ohba, K., Ikeuchi, K.: Detectability, uniqueness, and reliability of eigen windows for stable verification of partially occluded objects. PAMI **19**(9), 1043–1047 (1997)

19. Okada, R.: Discriminative generalized hough transform for object dectection. In: ICCV, pp. 2000–2005 (2009)

20. Ommer, B., Malik, J.: Multi-scale object detection by clustering lines. In: ICCV, pp. 484–491 (2009)

21. Tomasi, C., Kanade, T.: Detection and Tracking of Point Features. Tech. rep., Carnegie Mellon University (1991)

22. Tuzel, O., Porikli, F., Meer, P.: Region covariance: A fast descriptor for detection and classification. In: ECCV, vol. II, pp. 589–600 (2006)

23. Wertheimer, M.: Laws of organization in perceptual forms (partial translation). In: Ellis, W.B. (ed) A Sourcebook of Gestalt Psychology, pp. 71–88. Harcourt, Brace (1938)

24. Wojek, C., Roth, S., Schindler, K., Schiele, B.: Monocular 3D scene modeling and inference: Understanding multi-object traffic scenes. In: ECCV. vol. IV, pp. 467–481 (2010)

25. Yang, L., Zheng, N., Chen, M., Yang, Y., Yang, J.: Categorization of multiple objects in a scene without semantic segmentation. In: ACCV (2009)

26. Yarlagadda, P., Monroy, A., Ommer, B.: Voting by grouping dependent parts. In: ECCV, vol. V, pp. 197–210 (2010)

27. Yeh, T., Lee, J., Darrell, T.: Fast concurrent object localization and recognition. In: CVPR, pp. 280–287 (2009)

Chapter 7

Bluetooth in Intelligent Transportation Systems: A Survey

M. R. Friesen[1], R. D. McLeod[2]

[1]Design Engineering, University of Manitoba, Winnipeg, MB R3T 5V6, Canada

[2]Electrical and Computer Engineering, University of Manitoba, Winnipeg, MB R3T 5V6, Canada

ABSTRACT

The rise of Bluetooth-equipped devices in personal consumer electronics and in in-car systems has revealed the potential to develop Bluetooth sensor systems for applications in intelligent transportation systems. These applications may include measurements of traffic presence, density, and flow, as well as longitudinal and comparative traffic analysis. A basic Bluetooth sensor system for traffic monitoring consists of a Bluetooth probe device (s) that scans for other Bluetooth-enabled device (s) within its radio proximity, and then stores the data for future analysis and use. The scanned devices are typically on-board vehicular electronics and consumer devices carried by the driver and/or passengers which use Bluetooth communications, and which then reasonably proxy for the vehicle itself. This paper surveys the scope and evolution of these systems, with system attributes and design decisions illustrated via a reference design. The work provides motivation for continued development of non-invasive systems that leverage the existing communication infrastructure and consumer devices that incorporate short range communication technology like Bluetooth.

KEYWORDS

Bluetooth sensing; Traffic monitoring; Intelligent transportation systems

1 INTRODUCTION

Intelligent traffic systems (ITS) hold the promise to improve roadway congestion and transportation infrastructure management by capitalizing on information derived from traffic monitoring. The increasing requirement and public expectation for accurate vehicular traffic information to manage traffic flows has precipitated the deployment of large scale traffic monitoring infrastructures. Typically, this has included the use of inductive loop detectors, microwave sensors and relatively expensive video cameras.

On-board vehicular electronic devices as well as consumer electronic devices are emerging as an alternative traffic sensing modality to complement the existing traffic monitoring and management infrastructure. This evolving infrastructure provides has the benefit of providing cost-effective, real time traffic data by leveraging existing telecommunication infrastructure such as the cellular phone network.

This paper reviews the application of Bluetooth sensing in relation to ITS. It is a field that has seen very rapid evolution in the past 10 years, and will undoubtedly continue to evolve rapidly. A survey of the current state of the art can serve as a point of reference for both future and past works. In this paper, wireless sensor networks based on Bluetooth sensing are presented as a practical means of collecting a statistical representation of traffic density and flow. A basic system configuration consists of a Bluetooth probe device (s) that scans for other Bluetooth-enabled device (s) within its radio proximity, and then stores or forwards the data for future analysis and use. The scanned devices are typically on-board vehicular electronics and consumer devices carried by the driver and/or passengers which use Bluetooth communications, which reasonably proxy for the vehicle itself.

Thus, the data provide the information needed to extract a reasonable approximation of traffic presence, density, and flows.

The work reported here is contextualized within many of the uses initially suggested under the IntelliDrive initiative [1] that are oriented towards improving mobility within surface transportation systems. However, advanced applications are not practical to deploy on scale without leveraging implementations based on networked and web services and evolving internet technologies. Currently, web service realization for ITS applications is promising, since a uniform middleware can be achieved while utilizing the underlying network infrastructure [2]. The applications reviewed in this paper demonstrate this integration.

Early vehicular telematic applications were user-centric, whereas innovative applications that are now within reach often combine crowdsourced information with the objective of data collection and analysis for statistical reliability and generalizability [3]. These newer class applications are cohort-centric rather than individual- or user-centric. An early reference to the use of crowdsourcing for ITS [4] pursued the idea via a Smartphone app than from a wireless sensor network. These applications often rely on inferencing from GPS-equipped probes or floating cars, with the intent to capture the behaviours of a statistically significant portion of vehicles, such that meaningful inferences can be made and potentially generalized to the entire population of vehicles [5], [6]. Being statistical in nature, in some cases, only a very small amount of floating car or probe data are required to infer significant events such as congestion build-up or dissolution [7]. In addition to the work relative to estimating traffic from GPS-enabled probe devices and map services, an alternative includes the possibility of using drivers' cellphone trajectories as directly as proxies for vehicles, without an intermediary Smartphone app, as could be provided from a mobile cellular service provider [8–10]. Cellphone trajectories are typically coarser-grained in both space and time but can be overlaid on a traffic grid and allow for some degree of traffic flow inferencing [11].

This survey is directly related to the use of Bluetooth transceivers to crowdsource data from a statistically significant sample of vehicles, where the data can be analyzed for traffic reliable inferencing. Bluetooth is a short-range, wireless telecommunications standard that defines how mobile phones, computers, personal digital assistants, car radios, GPS units, and other digital devices can be easily interconnected. A prototypical example of the technology configuration is the interconnection of a driver's or passengers' mobile phone to a wireless earpiece of vehicle audio system for hands-free operation while driving.

The remainder of this paper is organized as follows. Section 2 provides a background of Bluetooth in Intelligent Transportation Systems (ITS) from some of the earliest academic references in the research area to the present. Section 3 discusses Bluetooth relative to the methods and data that can be readily obtained in a non-intrusive manner in an ITS context. Section 4 overviews the design considerations within a Bluetooth configuration for ITS applications, and Section 5 provides a provides a an illustrative reference design which encompasses many of the phenotypes of a typical Bluetooth sensor network for ITS and illustrates the ease with which data can be collected, stored and presented.

2 A SURVEY OF BLUETOOTH IN ITS

The earliest references to using Bluetooth for tracking purposes were typically unrelated to vehicular traffic and ITS. Within overall objectives of safety and flow monitoring, early examples included Bluetooth tracking systems to track of children at a zoo [12] and students at a University [13]. The realization that wireless sensor networks generally may play a significant role in traffic monitoring emerged in the literature in the mid-2000s [14] without explicitly mentioning Bluetooth.

Bluetooth may now appear to be an obvious methodology for non-intrusive traffic detection and estimation; however, industry reports and tests as late as 2010 evaluated various traffic sensors without considering Bluetooth as an option [15]. In a 2010 study, the authors almost apologetically implied that limiting their system to Bluetooth networks was a potential limitation, and indicated that their work could be applied to Wi-Fi devices as well [16]. The field since then has borne out that the Bluetooth devices serve as better proxies for vehicles than Wi-Fi devices would have been.

Another concurrent stream of research has been associated with the role of GPS in traffic monitoring relative to estimating travel times, vehicle density, and vehicle flows. In a more ambitious program where data was collected from GPS-enabled cellular devices, the work indicated that a 2-3 % penetration of GPS enabled cell phones in the driver population was sufficient to provide accurate measurement of the velocity of traffic flow [6]. Although the percentage of GPS enabled cell phones is clearly above the 2-3 % of the population, it is also a requirement that the GPS unit be physically on and driver data voluntarily be provided (either implicitly or explicitly). In comparison to Bluetooth, the only requirement is that the probed device has its Bluetooth enabled or discoverable, which is more often the case than having GPS enabled as well.

The potential of Bluetooth in traffic monitoring began to appear in the academic literature near 2010 [16, 17] although a small number of early field trials by local government transportation departments and agencies date to as early as 2008 [18] and 2010 [19]. Another early reference to Bluetooth sensing in a 2009 academic thesis emphasized optimal sensor location as opposed to data collection [20]. These were some of the first publications to consider Bluetooth as a means of collecting data for traffic monitoring and ITS management. A potential exception may be a 2004 reference to Bluetooth for ITS where Bluetooth is considered as a means of inter-vehicle communication as opposed the use of Bluetooth as a traffic monitoring sensor [21].

The reliability of Bluetooth sensor systems for ITS depends on widespread Bluetooth-enabled device penetration. In a study of traffic through the Limfjord Tunnel, Bluetooth penetration was estimated at between 27 and 29 % [22]. This appears to be considerably higher than many other reports at or about the same time frame. In a longitudinal study, Bluetooth penetration was seen to increase dramatically with the number of unique MAC addresses seen increasing over a year by 26 % [22]. Rationale for the increased penetration was assumed to be the popularity of GPS units combined with an increasing number of vehicles with built-in Bluetooth. This reported level of Bluetooth penetration augers well for the continued investigation into these technologies and increased levels of statistical confidence as an increasing number of Bluetooth devices become discoverable.

Most often, one thinks of the Bluetooth-enabled cellular phone as a foundational component in a Bluetooth sensor system for ITS. This premise is supported by cellular penetration rates in Canada and the U.S as well as elsewhere. In the first quarter of 2012, Canada had 28 million cellular subscribers in total, which is a penetration rate of up to 80 % if no duplicate devices to a single subscriber are considered. While not all cellular phones are Bluetooth-capable, Nielson reports Smartphone penetration of up to 64 % of mobile phone owners in the U.S. by August 2013 [23].

In one of the earliest references, researchers in 2008 envisioned the basic system components of a Bluetooth sensor system for ITS that have since evolved, stating that “one could easily imagine a battery-powered, Bluetooth enabled, smart cell phone in a plastic case chained to the side of the road to collect much more substantive travel time estimates over 24 h or 7 d to much more precisely characterize operational characteristics of either a signalized corridor or a construction work zone. Those data might be logged for later download" [18].

In general, these early studies typically focused on applications to vehicle travel times estimates (including travel time delays due to roadway obstructions) and origin–destination estimates on urban arterials and freeways. Potential applications to network analysis (shortest path), congestion reporting, bicycle and pedestrian travel times, and before-after studies were also speculated [19]. Other research interests included the quality of data produced by the Bluetooth detection of mobile devices for applications to travel time forecasting and estimates of time-dependent origin–destination matrices within an Advanced Traffic Information System (ATIS) that supplies information to drivers [24]. More recent studies likewise investigate the role of Bluetooth sensors to estimate travel times [25, 26] and vehicle velocities [27].

Privacy considerations were also foreseen early on, with the recommendation that organizations that implement Bluetooth-based tracking for ITS applications develop practices that encrypt and preclude storing MAC addresses for more than a few hours [18].

Early studies also observed asymmetry of traffic data collected in opposite directions (e.g. westbound vs. eastbound) and attributed this to antenna position. Others extended this line of investigation, characterizing antenna patterns and observing that an omnidirectional radiation pattern is most suitable for Bluetooth data collection [28]. This is not unexpected, as it reduces the complexity of antenna placement and analysis when implementing a larger system of Bluetooth sensors.

In one such study of antenna characterization relative to travel time data collection using Bluetooth, the proportion of unique Bluetooth MAC addresses read relative to the known total traffic flow was reported to be 10 % [28]. Although published in 2013, these data appear to have been collected in 2011 and as such, they represent a significant increase in the Bluetooth MAC address reads reported just three years earlier at 1 % [18]. This improvement is likely due to both higher gain antennas developed combined with the precipitous increase in the number of Bluetooth devices in vehicles over that same time period. Currently, the collecting of CoD information is largely absent in the available literature.

Furthermore, the role that Received Signal Strength Indicator (RSSI) may play in the Bluetooth data collection process in ITS arose as an explicit research question around 2011 [29], with little other apparent work on investigating RSSI as a Bluetooth data source specifically for vehicular traffic. One of the few studies available explores using RSSI as means of estimating distance from the scanning point and using Class of Device information as a potential means of differentiating pedestrians from vehicles [30].

System calibration and validation studies are becoming more rigorous as more Bluetooth sensor systems for ITS are deployed. Opportunities exist to calibrate the data with existing traffic measurement devices, including loop-detectors, mechanical counters, travel survey data of various forms, and increasingly, data inferred from cellphone trajectories. In one novel study, Bluetooth sensor data has been compared with automatic license plate recognition (ALPR) systems [17]. If available, ALPR provides as nice alternative for validation as both ALPR and Bluetooth sensors also label the data, in contrast to loop-detectors, counters, and survey data. Systems have been developed that employ a variety of sensors and their fusion in scaling up systems [31], such as the fusion of loop based detection with Bluetooth-enabled devices [32]. While Bluetooth devices appear to have been limited to defined probe vehicles resulting in a low quantity of Bluetooth data, the work demonstrated opportunities in multi-sensor data fusion using Bluetooth data in conjunction with auxiliary measurements. Another instructive example of multi-system calibration potential is with the Anonymous Wireless Address Matching (AWAM) proof of concept demonstration with the City of Houston on an urban arterial [33]. Travel times and speeds collected on identical roadway segments using a probe-based toll tag (AVI) system and the AWAM system were compared with excellent correlation.

While large-scale deployment of commercial Bluetooth traffic monitoring is still in its infancy, a number of pilot programs of various scales are being deployed. In Clark County, a pilot program costing $540,000 has approximately 20 Bluetooth probes installed collecting data along the Andresen corridor which experiences relatively high traffic, in an effort to determine whether the system can provide the information that traffic engineers need [34]. The study reports the system reading 3-5 % percent of vehicular traffic via Bluetooth MAC addresses, and they have recognized this to be sufficient to provide information on traffic flow. According to the authors, sufficient data and analysis is anticipated by early 2014 for agencies to use the outcomes to adjust traffic signal timing. Similar pilot programs are ongoing in many other countries [35–37]. While many pilot programs have focused on traffic flow on corridors and arterials, there are other Bluetooth scanning applications that are associated with work zone diversions [19, 38, 39]. At present, the majority of pilot programs are related to travel time informatics and Bluetooth data assessment towards building evidence-based cases for traffic signal retiming [40].

One of the earliest commercial system that used Bluetooth for vehicle identification for travel time estimation appears to be BLIDS [41] (http://www.blids.cc/). BLIDS was introduced early 2008 with over 50 systems deployed primarily along corridors. Traffax Inc. (http://www.traffaxinc.com/) is also one of the early commercial vendors of Bluetooth traffic monitoring systems, with the system known by the trade name BluFax and introduced in 2009. Traffax has a patent application pending (CA2711278 A1) which claims a priority date of January 2008, (provisional filing) but may come under some challenges as the system of [18] was already deployed at that time. Blipsystems (http://www.blipsystems.com/home/) supplies a commercial solution to Bluetooth tracking and traffic monitoring, with their product is denoted BlipTrack. Another commercial system is TrafficCast (http://trafficcast.com/) with a product denoted BlueTOAD. Iteris (http://www.iteris.com/) also has similar product denoted Vantage Velocity for capturing Bluetooth MAC data but does not appear to have integrated cellular connectivity. In these commercial systems, most if not all of the components discussed in this survey are included as product offerings. Somewhere in-between commercial systems and academic prototypes, various transportation institutes and agencies are leveraging intellectual property developed at or in conjunction with universities. An excellent example is the Texas A&M Transportation Institute [37, 42].

With the observed data gathering potential of Bluetooth for ITS applications combined with the emergence of a 'big data' culture, numerous references to systems and implementations have appeared from industry, governments, and academia. This diversity of approaches, motivations, and originators of the research lends credibility to the expanding role that Bluetooth sensing can play in ITS. In spite of this, there continues to be a need to explore system integration and validation for the various combinations of system configurations that can be envisioned.

3 BLUETOOTH TECHNOLOGY

Bluetooth is one of several available wireless technologies that may be employed to assist in resolving location extracted from a consumer electronic device (Table I). This survey focuses on classical Bluetooth 2.0 with a range that is well-tailored for monitoring or detecting devices residing in or integrated into vehicles, such as Smartphones, Bluetooth earpieces and car audio. NFC and BLE 4.0 are more recent market entries with emphasis on low power and more personal communication or very body-centric networking. WiFi and cellular are intended for wider area networking.

Table I. Various wireless proximities

Technology	Range
NFC (RFID)	~ 20 cm
BLE 4.0	<25 m (estimated)
Bluetooth	1-100 (class dependent)
WiFi	5-100 m (typical)
Cellular	km+(cell sector –typical)

In addition to Table I, there is a variety of wireless networking technologies developed over the IEEE 802.15.4 protocol such as ZigBee or XBee. These and similar are not considered here as they have not established a dominant presence in consumer devices such as smartphones. However, they are technologies to stay aware of as automobile manufacturers incorporate greater degrees of low cost wireless technologies into product lines. Relative to the basic functionality of inferencing vehicle presence by scavenging radio signals, ZigBee could provide an alternative to Bluetooth, although at this time is difficult to foresee ZigBee to be as pervasive as Bluetooth.

In general, Bluetooth scanning requires a device to probe the local wireless environment and detect the proximity of Bluetooth radios. In the ITS context, the proximate Bluetooth radios (typically, drivers' or passengers' Smartphones, earpieces, or on-board car audio) detected by the probe device serve as proxies for vehicles. The objective of the probe devices (aka probes) is to collect information on vehicle presence (detection by one probe) and vehicle trajectory information (detection by multiple probes in sequence) via Bluetooth device discovery, and then transmit this information via either an intermediate wireless tier or directly over a cellular network to a web service or backend server. The communication protocols between probes and backend servers are typically based on the TCP/IP protocol stack, leveraging not only the physical communication infrastructure but also the highly developed Internet IP infrastructure.

The data collected from a Bluetooth scan can be fairly detailed, including a detected Medium Access Control (MAC) address, Class of Device (CoD) information, as well as metadata from the manufacturer or as specified by the device owner (Figure 1). While getting all of this information from a device and storing it could lead to security and privacy concerns, the MAC address, the primary method of identifying a Bluetooth device, can easily be anonymized in a fashion such that the device can be uniquely identifiable from other devices while preserving a user's identity to a certain degree.

id	date	agent	agentmac	device	devicemac
58983	2010-10-25 12:38:31	Agent3	F40B93BDCC7C	BlackBerry 8520	307C30FB6AA7
34643	2010-10-18 09:24:32	Agent3	F40B93BDCC7C	Ameena Bajer-Koulack?s MacBook	60334B22AE90
14583	2010-10-12 14:55:19	Agent1	F40B93C44389	Emmanuel Abrokwah?s MacBook Pro	C8BCC8B04EC1
25501	2010-10-13 13:09:16	Agent2	F40B93C3F258	Sam	001CA41852A9
31712	2010-10-15 13:29:36	Agent1	F40B93C44389	Kevin Baros-Novak?s MacBook Air	90840DF45EA4
26567	2010-10-14 08:04:23	Agent2	F40B93C3F258	HF99	0013E901FDED
42981	2010-10-20 09:39:46	Agent3	F40B93BDCC7C	Scott Vieira?s MacBook Pro	5860359E3698
10489	2010-10-12 09:18:50	Agent3	F40B93BDCC7C	FUJITSU-LAPTOP	00037A875141
3636	2010-10-08 16:03:14	Agent2	F40B93C44389	Parrot v5.11C	00121CECD098
1167	2010-10-07 15:58:59	Agent2	F40B93C44389	General Motors	001EB2DCDC31
56109	2010-10-21 13:39:46	Agent1	F40B93C44389	Jupa	C8BCC8DAAB76
7556	2010-10-09 16:27:09	Agent3	F40B93BDCC7C	C510a	0024EFC4DC6F
5904	2010-10-08 16:59:06	Agent3	F40B93BDCC7C	0000	0017D57473C6
22040	2010-10-13 09:24:42	Agent3	F40B93BDCC7C	hamster	001F5B6EB5B3
3923	2010-09-18 02:36:30	Agent2	F40B93C3F258	emmaddaford<3	0026E2EBF504
9358	2010-10-10 15:18:52	Agent3	F40B93BDCC7C	SPH-M610	001D25EAE5A8
30855	2010-10-14 16:27:25	Agent1	F40B93C44389	Ronnie Araya MacBook Pro	002608D8C018
32937	2010-10-16 13:50:13	Agent1	F40B93C44389	M350	001D256A1752
473	2010-10-06 16:17:46	Agent1	F40B93C44389	MYRNA	0012D2C6C1E8

Figure 1. Simple Bluetooth Probe Scan Data

The Bluetooth address itself is a unique 48bit device identifier, where the first 3 bytes of the address are assigned to a specific device manufacturer by the IEEE (www.ieee.org/), and the last 3 bytes are freely allocated by the manufacturer. Even if the manufacturer of a device is known, the number of possible Bluetooth addresses is immediately limited to 16,777,216. As well, since only devices that are in discoverable mode can be detected by the scan, a user can simply turn their device off (non-discoverable) to avoid being detected. There are software tools available which allow brute-force discovery of non-discoverable devices, an early example of which is RedFang, but this is usually too complex for a minimal hardware configuration and not necessary for data collection purposes associated with ITS. It is also possible to burrow deeper into Bluetooth connections to provide connection-based tracking for fine granular or building-wide device tracking [43].

It is desirable, but not necessary, to employ probes in such a way that the system is scalable. This is one of the most unique features of a Bluetooth data collection system. An organization can start with a very limited and even a portable system and easily scale it to meet increased sensing demands. In this regard, the backend data collection, storage and analysis hardware are conservative. Modern web servers and connectivity are more than sufficient to collect the volume of data that could be collected even from a large number of Bluetooth probes. The scalability issue has essentially been further resolved by existing cellular and internet infrastructure, web servers, and data services collectively being developed as a Service-Oriented Architecture (SOA) [44, 45].

The primary equipment used as the Bluetooth probe device must have the ability to be set for device discovery. Several of the existing Bluetooth modules libraries are designed to detect eight devices per inquiry, which is insufficient for most vehicular traffic applications. The limitation of eight appears to be a consequence of the anticipated use cases of Bluetooth enabled consumer devices, where supporting eight connections is likely more than sufficient. In effect, many existing Bluetooth modules are designed for connecting with the Bluetooth devices it discovers, whereas for ITS Bluetooth data collection, the primary requirement is only for devices to be detected. During traffic congestion periods, it is desirable to discover as many surrounding devices as possible, but a connection to the surrounding devices is not required. The important aspect is to ensure that the selected Bluetooth probe modules can discover as many devices as possible. In the reference design (Section 5), up to 250 unique devices per inquiry were detectable and hence recordable. This detection capability is not a restriction of the standard but rather particular implementations. For example, a detailed description of the Bluetooth discovery protocol that simulates the detection of 15 devices within a few seconds is available in the literature [46].

4 BLUETOOTH SENSOR SYSTEM: ATTRIBUTES & DESIGN DECISIONS

When designing a Bluetooth sensor system for ITS applications, there are choices in system attributes that become design decisions unique to the context and objectives of the system in deployment (Figure 2). The basic configuration requires the designer to decide what type of probe device (s) will be used, how many probe device (s) are required, and where and how they will be located and fixed in the environment.

- Commercial/Prototype: In academia, a prototype system can often be easily assembled for several thousand dollars and provide entry-level data for exploring Bluetooth device detection. A commercial-grade installation of similar scope would likely be an order of magnitude greater in cost. Intermediate-grade systems are also available, for example [30].

- Fixed/Portable: A fixed system implies a permanent or semi-permanent installation, whereas a portable system implies a system set-up or tear-down time of several hours maximum, and a deployment period measured in days, weeks, or months. In the case of a portable system, consideration has to be application domain, and a portable system offers the convenience of being more easily redeployed. A portable system typically implies a storage battery and possibly a battery charging system, such as solar. A fixed system is typically more robust but greater attention in sensor placement is required as the initial placement decisions may not be easily changeable. Fixed system may also rely on battery power, although a permanent power supply may be cost-effective as well. In both fixed and portable systems, GPS positioning is required. Both fixed and portable systems may be online or offline

- Online/Offline: An on-line system has the potential for real-time data collection (and potentially, analysis), by data being stored in the probe for only very short periods of time and backhauled at regular intervals over a wireless connection to a dedicated server. In an offline system, a requirement exists for much more significant data storage and relatively simple retrieval. A minimum requirement would be for the probes to write data to an onboard SD card or similar, and the manual data retrieval protocol to be planned. On-board data storage is typically Flash-based and very robust.

- Wireline power/battery/solar: The choice of power to the probe devices depends largely on the intended application. A portable application will most probably be battery powered, and depending on it intended duration and the environment, it may also be equipped with a solar charging system. Wireline power may be an economical alternative in a fixed system, although a fixed system may also run on battery power.
- Networking: Networking considerations are typically limited to online systems, where the designer must consider means of data transport. Two of the more likely networking technologies are cellular (e.g. GSM), or WiFi. The consideration in selecting WiFi would be to ensure adequate coverage over a wide area.
- Tiered wireless configuration: A tiered wireless sensor network is a very common architecture for sensor networks (e.g. [30] and the reference design in Section 5). In a system with multiple probe devices, a design decision must be made whether each probe will be equipped with its own WiFi or GSM/GPRS module, allowing direct communication from each probe to the back-end server. The alternate would be to implement a middle wireless tier, adding some complexity to the system. An example, of a multi- tiered wireless implementation is presented in Section 5. It is these authors opinion however, that a tired wireless system should be avoided and direct communication to the probes be supported.
- RSSI capable: Only a very limited number of published works incorporated RSSI for traffic monitoring in an ITS context [29,30, 47]. At time of writing, RSSI data does not play a significant role in Bluetooth systems for ITS applications, although the rapid evolution of this area in general may illuminate the potential and utility of RSSI data within a few years.
- Bidirectional: A bidirectional system may be able to proactively communicate over Bluetooth to 'subscribers' or discovered devices, for example by providing traffic alerts. For example, a potential business model would allow users to purchase low-cost devices solely for the purpose of being tracked [18], with a subsequent feature of being able

to backcast from the ITS to the device. This device may be as redundant as a Smartphone running a suitable app. The opportunity to also log a subscriber's OBD-II data automatically would also require bidirectional communication via an established connection with a simple Bluetooth OBD-II dongle.

- Remote monitoring and/or control: The Bluetooth probes should be chosen to be remotely monitored and preferably also configurable. The physical environment of installation sites (heavy traffic areas, exposure to all weather conditions) may make on-site monitoring and configuration both uncomfortable and potential dangerous. Remote monitoring allows for early detection of malfunctioning probes and other inconsistencies. Remote configuration can range from configuring the sampling rate and/or sleeping sensors when not required, which becomes critical for battery powered units. The implication of remote monitoring and configuration is that wireless access via WiFi or cellular is available to the probes. .
- Cross validation/calibrating: As a Bluetooth sensor is sampling a proportion of the by-proxy vehicle population, by definition considerable emphasis has to be placed of validating and calibrating. A Bluetooth sensor network is relatively easy to install and provides considerable opportunity for sensor fusion. At minimum, designers must consider Bluetooth radio ranges of the probe (s) relative to the sampling area and clock synchronization between multiple probes. In data analysis, considerations include but are not limited to the ability to handle multiple MAC address reads from a given probe sensor that represent different vehicles as well as multiple reads of the same vehicle (e.g. a vehicle stopped at a traffic signal for several sampling periods), simultaneous MAC address reads from two or more probe sensors, one or more MAC address reads of a single MAC address by multiple sensors either simultaneously or in sequence, and MAC address reads of devices that are not necessarily sourced from a vehicle. The fact that the data is labeled can be used to provide some level of differentiation. There will always be some degree of uncertainty; for example, the data analysis is unlikely to be able to definitely differentiate a single vehicle from public transportation (e.g. a bus with 40 passengers and multiple Bluetooth reads). Managing uncertainty is one of the more academic aspects associated with

Bluetooth traffic monitoring and promises to be a rich area of research [48].

- Security: By virtue of the fact that all wireless communication devices need to signal to some degree in plain sight, security will always be an issue. It is not possible to alter this fact, as communicating devices need a standard means of identifying one another.

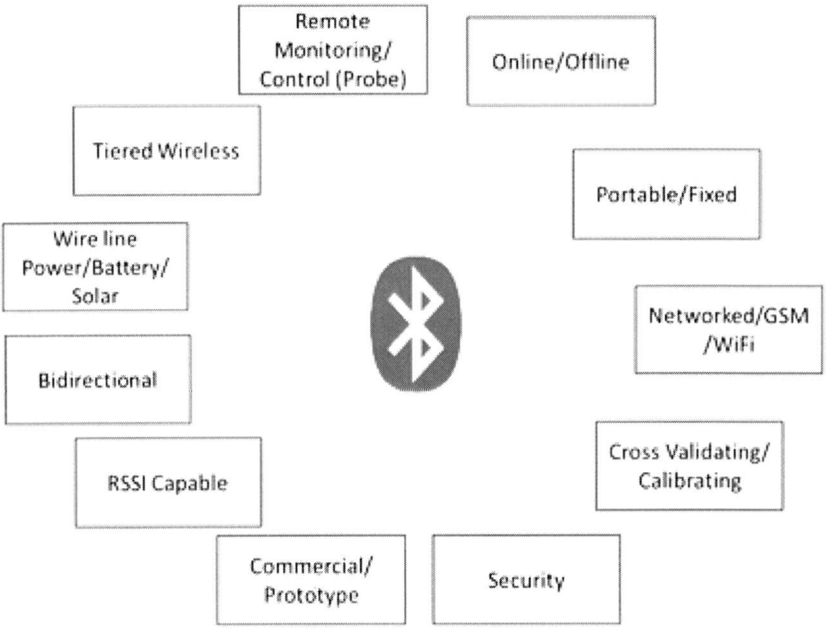

Figure 2. Bluetooth traffic monitoring design decisions

Bluetooth scanning for ITS applications such as simple traffic flow will become contextualized within big data concepts, in which the opportunities for exploring and exploiting the rich suite of data such systems are capable of generating is a significant research area in its own right. Examples include generating trajectories from uncertainties in measurement and detection [49] using techniques like Hidden Markov Models. Similar efforts will produce forecasting models that use massive amounts of real-time Bluetooth device data.

5 A REFERENCE DESIGN

This section presents a reference design as an example of one combination of many of the system attributes and design choices overviewed above. The system used multiple Bluetooth transceivers consisting of one master node or access point and multiple sensor probes around a major intersection in a Winnipeg, Canada, during winter 2013 (Figure 3) [50]. The multiple probes collected vehicle presence information (detection by one probe) and vehicle trajectory information (detection by multiple probes in sequence) via Bluetooth device discovery, and then transmitted this information to the master node via the 802.15.4 protocol.

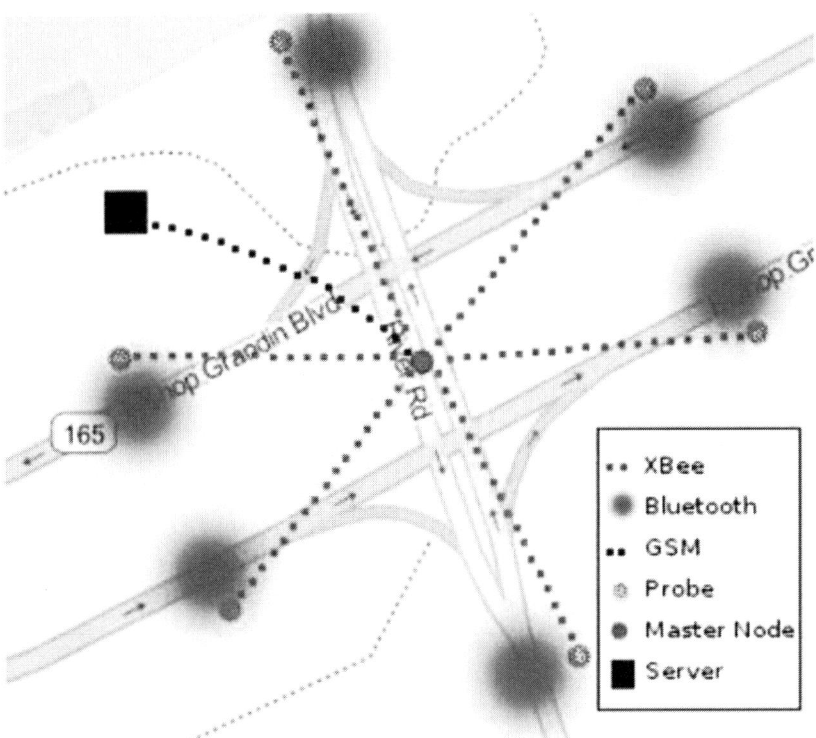

Figure 3. High level system overview

5.1 Architecture

The basic system architecture is that of a Bluetooth sensor network, interconnected with an XBee/802.15.4 middle tier to the master node, and then a GSM wireless backhaul tether to a web data collection and processing web portal. XBee Pro was selected as the middle tier wireless networking technology as it offers a low power solution with sufficient range for the interconnection between the master node and sensor probes. GSM was selected as the cellular tier as a means of aggregating and forwarding data collected to a web server for processing and display. The data sent to the central server displayed the current traffic density and average velocity of vehicles at the intersection on a web front end (including mobile website) at five-minute intervals.

5.2 Probe design

The probe design uses an Arduino Uno development board, which utilizes the ATMega328 microcontroller, an XBee module and a Bluetooth Pro module, both of which connect to the Arduino module (Figure 4).

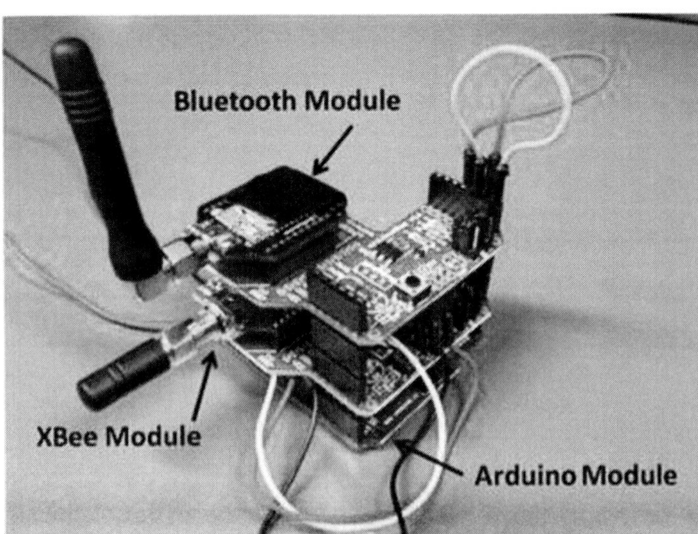

Figure 4. Probe node

During the data collection process at each probe, the information was organized in a consistent manner. Each device frame is 8 bytes and included a marker byte, a 4 byte timestamp, and 3 bytes for the truncated MAC address. To increase privacy, only half of the MAC address is recorded which still provides 16 million unique combinations for one probe network. The first 3 bytes of the packet are reserved bytes; a control byte, a length byte, and a probe id byte. The maximum packet size that can be produced is 100 bytes, as this is the size of the receive buffer of the XBee module.

The XBee module on the master node was set as the network coordinator, and the XBee module on each probe was set to associate with the network coordinator. To save power, the XBee module on the probes were configured to hibernate when not in use. Due to changes in temperature and other internal and external factors on the probes and master node, the calculated time offset at various nodes slowly drifted over time, requiring scheduled clock resynchronization.

The function of the probe is illustrated in the main flow diagram of Figure 5, and the scan flow sub diagram is illustrated in Figure 6.

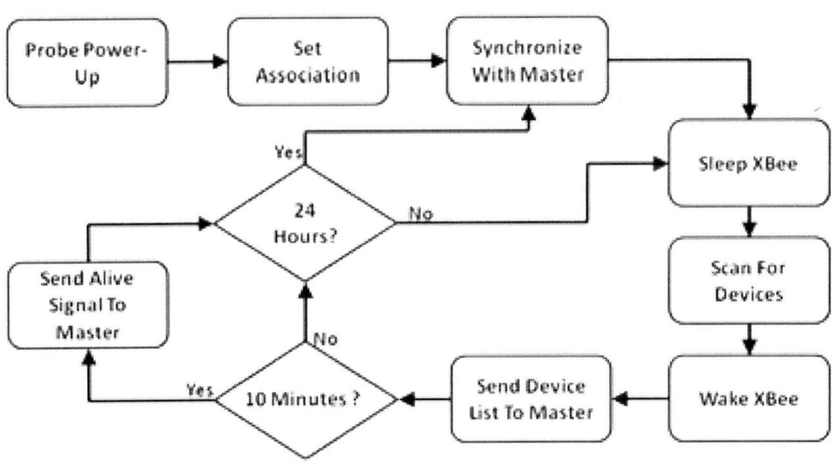

Figure 5. Probe flow diagram

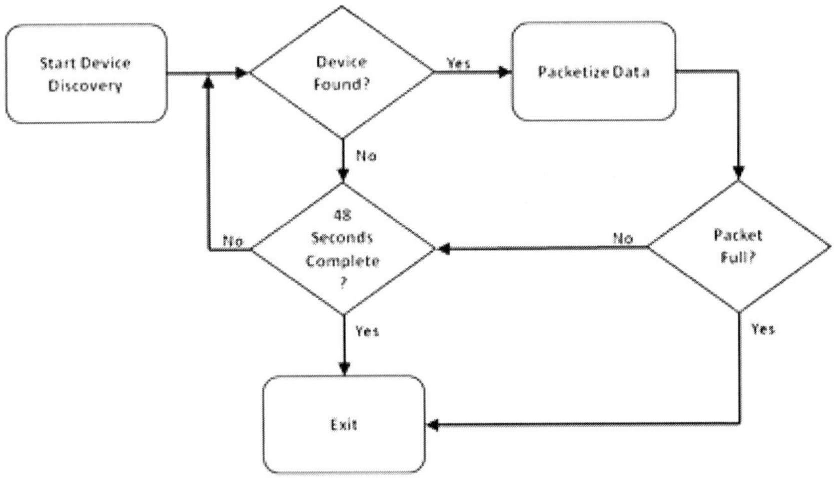

Figure 6. Bluetooth scan diagram

5.3 Master node

The master node consisted of Arduino-based GBoard from iteadstudio.com. For the GSM module, a preexisting library was used, with modifications and additional functionality to meet system requirements.

5.4 Power

The desired minimum run time of each probe was 24 h. A combination of battery-only and battery-plus-solar were used, demonstration the viability of a solar rechargeable source for the wireless sensor network. The rechargeable power sources were two 4,400 mA, 3.7 V lithium ion batteries hooked up in parallel to a step-up converter (to 7 V). Both batteries are connected to lithium-ion chargers powered by a 3.7 W, 6 V solar panel. A custom step-up converter was also designed to have lower current draw compared to the off the shelf step-up converter; thereby increasing the life expectancy of the supply.

5.5 Database and front-end

A MySQL database was used in order to store data, refreshed at five minute intervals, and display data to a user through a website or a smartphone app. Initial scripts were written where device MACs are compared to a table of existing MACs to find matching devices with a different probe location. When found, a probe-pair was created using the start probe, end probe and detection time difference; this was used to estimate the velocity of the device. Another PHP script was created that automatically runs every five minutes, refreshing the traffic data that is displayed on the websites. The data is stored as the total travel time for all vehicles and the total vehicle count through each probe-pair.

5.6 Implementation

A Bluetooth discovery test was conducted to ensure devices traveling in excess of 80 km/h could be detected. The XBee distance test consisted of testing the connectivity while increasing the separation between master and slave devices. Line of sight difficulties were encountered for distances exceeding 200 m. As a complete test of the system, four probes were placed along a major thoroughfare to capture the traffic over several 24 h timeframes for multiple iterations representative of different traffic and weather conditions. Probes and master nodes were mounted to light standards along the thoroughfare using steel strapping at an elevation of approximately 2.5 m. Several single-probe and multi-probe trials were carried out over winter, 2013. Data were cross-referenced to data obtained by mechanical traffic counters and to cellular service provider data that can serve as proxies for users' movements between cell towers. Qualitatively similar trends in traffic density and flow were observed from the data sources. The reference design was found to capture an average of 4.5 % of real vehicle traffic (when compared to mechanical counters), which is above the 2-3 % conjectured as being required for statistically accurate traffic flower inferencing [6]. Initial findings lent credibility to the reference design as having the characteristics of a viable full-scale system. The cost of the reference design described above was approximately $1,500 in 2012 and 2013.

The reference system described above provides insights into the type of data that can be easily and cost effectively collected at a relatively inexpensive capital cost. The reference design, which was an academic prototype, is validated conceptually by others who report similar systems and investigations. In a study on traffic monitoring with Bluetooth sensors over ZigBee, an intervening multitier network is discussed [30]. Although a different protocol is deployed, the basic ideas and rationales are similar. Others similarly discuss the use of Bluetooth data to infer vehicle proximity as a means of estimating traffic characteristics, including the use of solar power to meet system energy requirements [51, 52].

6 ADDITIONAL CONSIDERATIONS

As of time of writing, the emerging wireless technology for inferring traffic is that based on Bluetooth 2.x. Bluetooth 2.x is the most widely deployed mid-range communication protocol has a range commensurate with distances typical of traffic and traffic control systems. The ubiquity of Bluetooth 2.x and its ready application to traffic-related contexts currently makes it a natural choice for ITS applications. In the future, communication alternatives and superior wireless versions - ostensibly developed for different user applications - will undoubtedly contribute to the data collected for ITS. For example, Bluetooth 4.0 is currently available for 'coin-cell-powered' devices and is currently not obviously applicable to common ITS applications; however, this applicability is likely to develop as new opportunities emerge. The same is true for other technologies, in the spirit that the most effective uses of a technology often emerge after its development rather than as a bounded or fixed a priori specification.

From our experience, one significant recommendation in this field of Bluetooth sensor systems for ITS would be to avoid ad hoc intermediate networks. The difficulties with installation and reliability make data collection networks that rely on lines of sight for communication too constraining. In addition, the XBee communications used in the reference design were power-intensive and not tuned for power conservation. As an alternative, each sensor should ideally be equipped with GSM/GPRS and have the backend service parse data from each probe directly. The additional costs of this approach are likely to be outweighed by the benefits accrued through much simpler deployment and lower maintenance. The intervening XBee, communication network also added complexity through the use of a less efficient protocol for data transmission than GSM directly. This follows the principle of "Lex Parsimonia" where the simplest system is most likely the best.

Future work should investigate the integration of mobile Bluetooth probes as well as stationary probes. This increases processing and analysis of the data collected, but would augment data collected by the stationary Bluetooth probes. An example of a vehicle trajectory while scanning, is in [53], while a patent application for a mobile probe can be found in [54]. Utilizing mobile probes would also allow for augmenting of the data with information such as acceleration at intersections, and would be a direct means of inferring environmental conditions such as ice and snow.

Finally, a decided advantage of a Bluetooth traffic monitoring system through a Services Oriented Architecture is that it can easily and inexpensively augment any existing traffic measurement system.

The real academic, engineering, and organizational challenges will lie in full-scale deployment of a Bluetooth sensor system at many intersections across a large urban centre. The probe sensor network may be augmented by a large number of probe vehicles that could also be configured to upload data from proximate devices detected while in transit. The only requirement for this additional data from probe vehicles is to augment the discovered Bluetooth MAC address and timestamp with GPS information. Once scaled to city-wide deployment, the data mining challenges will be considerable and will require a whole new big data approach, but will also become an invaluable input to an ITS.

ACKNOWLEDGMENTS

The authors thank the City of Winnipeg Public Works Department for financial support, and in particular Mr. Doug Hurl and Mr. Luis Escobar. In addition, the financial support of NSERC and NSERC NCE Auto 21 is gratefully acknowledged.

REFERENCES

1. Research and Innovative Technology Administration (2010). Achieving the Vision: From VII to Intellidrive. [Online]. Available: http://www.its.dot.gov/press/2010/vii2intellidrive.htm

2. R. Sengupta, C. Manasseh and S. Levenson,. "Service oriented architectures for VII – Extending web services to the roadway". [Online]. Available: http://www.its.berkeley.edu/volvocenter/VREF/UCBSOA_VIIChristian.pdf

3. E. Hossain, G. Chow, V.C.M. Leung, R.D. McLeod, J. Mišić, V.W.S. Wong, and O. Yang, "Vehicular telematics over heterogeneous wireless networks: A survey", Computer Communications, Volume 33 (Issue 7, 3), Pages 775–793 (May 2010)

4. J. Sorensen, "One in Evert Crowd", Traffic Technology International, February-March, http://www.dagstuhl.de/Materials/Files/10/10121/10121.XuBo1.Paper.pdf (2010)

5. B.S. Kerner, C. Demir, R.G. Herrtwich, S.L. Klenov, H. Rehborn, M. Aleksic and A. Haug, "Traffic state detection with floating car data in road networks. Proceedings of the 8th International IEEE Conference on Intelligent Transportation Systems, 44–49 (2005)

6. Herrera, J.C., Work, D.B., Herring, R., Ban, X., Jacobson, Q., Bayen, A.M.: Evaluation of traffic data obtained via GPS-enabled mobile phones: The Mobile Century field experiment". Transportation Research Part C: Emerging Technologies **18**(4), 568–583 (2010)

7. G. Halbritter, T. Fleischer, C. Kupsch, J. Kloas and U. Voigt, "Nationale Innovationsstrategien für neue Techniken und Dienste zur Erreichung einer „nachhaltigen Entwicklung" im Verkehr (Report FZKA 7157). Karlsruhe, Germany: Forschungszentrum Karlsruhe GmbH, (2005)

8. Y. Yim and R. Cayford, "Investigation of vehicles as probes using global positioning system and cellular phone tracking: Field operational test", Number UCB-ITS-PWP-2001-9, California PATH Program, Institute of Transportation Studies, University of California, Berkeley, CA, (2001)

9. J.L. Yagnace, C. Drane, Y.B. Yim, and R. de Lacvivier, "Travel time estimation on the San Francisco Bay area network using cellular phones as probes". Number UCB-ITSPWP-2000-18, California PATH Program, Institute of Transportation Studies, University of California, Berkeley, CA, (2000)

10. Alessandri, R. Bolla, and M. Repetto, "Estimation of freeway traffic variables using information from mobile phones", In Proc. American

Control Conference the 2003, volume 5, pages 4089–4094, Denver, CO, (June 2003)

11. R. Neighbour, R.D. McLeod, M.R. Friesen, and S.N. Mukhi, "Vehicular Traffic Modeling Governed by Cellular Phone Trajectories", VTC-Fall (2012)

12. BlueTags to install world's first Bluetooth tracking system, 2003. http://www.geekzone.co.nz/content.asp?contentid=1070.

13. M. Haase and M. Handy. BlueTrack – Imperceptible tracking of bluetooth devices. In Ubicomp Poster Proceedings, (2004)

14. Y. Wen, J.L. Pan and J.F.Le, "Survey On Application Of Wireless Sensor Networks For Traffic Monitoring", International Conference on Transportation Engineering 2007 (ICTE 2007)

15. E. Minge, J. Kotzenmacher, and S. Peterson "Evaluation of Non-Intrusive Technologies for Traffic Detection", Mn. DoT., Final Report #2010-36, http://www.miovision.com/wp-content/uploads/Evaluation_of_Non-Intrusive_Traffic_Detection_Technologies_pdf (September2010)

16. Barceló, J., Montero, L., Marqués, L., Carmona, C.: Travel Time Forecasting and Dynamic Origin–destination Estimation for Freeways Based on Bluetooth Traffic Monitoring". J. Transp. Res. Board **2175**, 19–27 (2010)

17. Y. Wang, Y. Malinovskiy, Y-J. Wu and U-K. Lee, "Field Experiments with Bluetooth Sensors", n.d., http://www.westernstatesforum.org/Documents/UW_FINAL_Bluetooth_WSRTF_6-17-10.pdf (2010)

18. Wasson, J.S., Sturdevant, J.R., Bullock, D.M.: Real-Time Travel Time Estimates Using MAC Address Matching. Institute of Transportation Engineers Journal, ITE **78**(6), 20–23 (2008)

19. R. Haseman, J.S.Wasson and D.M. Bullock, D M., Real-Time Measurement of Travel Time Delay in Work Zones and Evaluation Metrics Using Bluetooth Probe Tracking. Transportation Research Record: Journal of the Transportation Research Board, pp. 40–53, (2010)

20. M. Asudeqi, “Optimal Number and Location of Bluetooth Sensors for Travel Time Data Collection in Networks”, M.Sc. Thesis, University of Maryland

21. H. Sawant, J. Tan, Q. Yang, and Q. Wang, “Using Bluetooth and sensor networks for intelligent transportation systems,” Intelligent Transportation Systems, 2004. Proceedings. The 7th International IEEE Conference on, vol., no., pp.767,772, 3–6

22. B. Araghi, K.S. Pederson, L.T. Christenson, R. Krishnan, and H. Lahrmann, “Accuracy of Travel Time Estimation using Bluetooth Technology: Case Study Limfjord Tunnel Aalborg”, ITS World Congress, (2012)

23. Nielson, “Smartphone Switch: Three Fourths of Recent Acquirers Chose Smartphones” Available: http://www.nielsen.com/us/en/newswire/2013/smartphone-switch--three-fourths-of-recent-acquirers-chose-smart.html

24. Barceló, J., Montero, L., Marqués, L., Carmon, C.: Travel Time Forecasting and Dynamic Origin–destination Estimation for Freeways Based on Bluetooth Traffic Monitoring”. J. Transp. Res. Board **2175**, 19–27 (2010)

25. J.D. Porter, D.S. Kim and M.E. Magana, “Wireless Data Collection System for Real-Time Arterial Travel Time Estimates”, OTREC 10–16, http://www.oregon.gov/ODOT/TD/TP_RES/docs/reports/2011/wirelessdata.pdf, (March 2011)

26. S.E. Young, “Bluetooth Traffic Detectors for Use as Permanently Installed Travel Time Instruments”, STMD -MDOT/SHA, http://roads.maryland.gov/OPR_Research/MD-12-SP909B4D-Bluetooth-Traffic-Detectors_Report.pdf, (Feb. 2012)

27. Anonymous Wireless Address Matching (AWAM) proof of concept demonstration, Texas Transportation Institute, http://ttihouston.tamu.edu/bluetooth/projects.aspx

28. Porter, J.D., Kim, D.S., Magaña, M.E., Poocharoen, P., Arriaga, C.A.G.: Antenna Characterization for Bluetooth-Based Travel Time Data Collection”. Journal of Intelligent Transportation Systems: Technology, Planning, and Operations **17**(2), 142–151 (2013).

29. J.D. Porter, D.S. Kim, M.E. Magaña, P. Poocharoen and C.A.G. Arriaga, "Antenna Characterization for Bluetooth-Based Travel Time Data Collection", http://www.westernstatesforum.net/Documents/2011/presentations/OSU_KimEtAl_Final_AntennaCharacterization.pdf.

30. Vehicle Traffic Monitoring Platform with Bluetooth Sensors over ZigBee, [Online]. Available: http://sensor-networks.org/index.php?page=1129929946

31. G.A. Klunder, Y. S. Chen, K. Zantema, and K. Friso. "Consistent Traffic Modeling and Calibration at Different Resolution Levels Based on Various Sensor Data for Real-time Traffic Management", 2nd International Conference on Models and Technologies for Intelligent Transportation Systems, 22–24, Leuven, Belgium (June, 2011)

32. Bachmann, C., Roorda, M.J., Abdulhai, B., Moshiri, B.: Fusing a Bluetooth Traffic Monitoring System With Loop Detector Data for Improved Freeway Traffic Speed Estimation. Journal of Intelligent Transportation Systems: Technology, Planning, and Operations **17**(2), 152–164 (2013).

33. E. Hidle, "Bluetooth can help local traffic flow", http://www.columbian.com/news/2013/oct/21/bluetooth-traffic-flow-clark-county/

34. S. Yucel, H. Tuydes-Yaman, O. Altintasi, M. Ozen, "Determination of Vehicular Travel Patterns in an Urban Location using Bluetooth Technology", https://itswc.confex.com/itswc/AM2013/webprogram/ExtendedAbstract/Paper11460/ITS_America_sule_yucel.pdf

35. M. Blogg, C. Semler, M. Hingorani, R. Troutbeck, "Travel Time And Origin–destination Data Collection Using Bluetooth MAC Address Readers", Australasian Transport Research Forum 2010 Proceedings, p. 1–15 (2010)

36. D.D. Puckett and M.J. Vickich, "Bluetooth-Based Travel Time/Speed Measuring Systems Development", Texas Transportation Institute, DTRT06-G-0044, http://utcm.tamu.edu/publications/final_reports/Puckett_09-00-17.pdf, (2010)

37. J. Effinger, A. J. Horowitz, Y. Liu, and J. Shaw, "Bluetooth Vehicle Reidentification for Analysis of Work Zone Diversion." In Transportation Research Board 92nd Annual Meeting, no. 13–2159. (2013)

38. S. Zinner, "A methodology for using bluetooth to measure real-time work zone travel time." PhD diss., Georgia Institute of Technology, 2012

39. C.M. Day, J. S. Wasson, T. M. Brennan, and D. M. Bullock. Application of Travel Time Information for Traffic Management. Publication FHWA/IN/JTRP-2012/06. Joint Transportation Research Program, Indiana Department of Transportation and Purdue University, Indiana

40. Steiner and B. Sick, "A new method for travel time estimation on long freeway sections", Institute of Data Analysis and Process Design, Zurich University of Applied Sciences. European Journal of Transport and

41. http://tti.tamu.edu/03/01/now-taking-it-to-the-streets-collecting-travel-time-data-speed-with-bluetooth-technology/#applying (2010)

42. S. Hay and R. Harle, "Bluetooth tracking without discoverability," Proc. of the 4th Int'l Symp. on Location and Context Awareness. Springer-Verlag, pp. 120–137 (2009)

43. H. Quintero, L. Felipe, F.M. Pérez, D.M. Jorquera, and Virgilio Gilart Iglesias. "SOA-based Model for the IT Integration into the Intelligent Transportation Systems." IEEE ITSC2010 Workshop on Emergent Cooperative Technologies in Intelligent Transportation Systems, (2010)

44. X. Tao, C. Jiang, Y. Han, "Applying SOA to intelligent transportation system", Proceedings of the IEEE International Conference on Services Computing (SCC'05), Vol.2, pp. 101–106 (2005)

45. G. Chakraborty, K. Naik, D. Chakraborty, N. Shiratori and D. Wei, "Analysis of the Bluetooth device discovery protocol. Wireless Networks", Wireless Networks A. Saeedi, S. Park, D.S. Kim, J.D. Porter, "Improving Accuracy and Precision of Travel Time Samples Collected at Signalized Arterial Roads with Bluetooth Sensors ', Volume 2380, pp. 90–98, (2013)

46. F. Zheng, "Modelling Urban Travel Times", Thesis Delft University of Technology, http://repository.tudelft654/Dissertation_Zheng.pdf (2011)

47. J.D. Lees-Miller, R.E. Wilson, and S. Box. "Hidden Markov Models for Vehicle Tracking with Bluetooth." In Transportation Research Board 92nd Annual Meeting, no. 13–3032. (2013)

48. M. Friesen, R. Jacob, P. Grestoni, T. Mailey and R.D. McLeod, "Vehicular Traffic Monitoring using Bluetooth", CCECE 2013

49. R. Cavaliere, P. Valleri, R. Kloibhofer, "The INTEGREEN approach for the assessment of environmental traffic management policies, 9th ITS European Congress, Dublin, Ireland, 4–7 June 2013

50. F. Biora, F. d'Aprile, and R. Marino, "A Large Scale Application for Bluetooth-based Travel Time Measurement in the Netherlands", 19th ITS World Congress, Vienna, Austria, 22 to 26 October (2012)

51. J. Benavides, B.C.P. Demianyk, S.N. Mukhi, M. Laskowski, M.R. Friesen and R.D. McLeod, "Smartphone Technologies for Social Network Data Generation and Infectious Disease Modeling", J. of Medical and Biological Engineering, 32(4): 235–244

52. L.M. Peterson, "Obtaining vehicle traffic information using mobile Bluetooth detectors", https://www.google.com/patents/ US20120276847, Google Patents (2012)

Chapter 8

A Comparative Study on Fuel Consumption Reduction Effects of Eco-Driving Instructions Strategies

Shuichi Matsumoto [1], Taehwi Park[2], Hironao Kawashima[3]

[1]Faculty of Science and Technology, Keio University, 3-14-1 Hiyoshi, Kohoku-ku, Yokohama 223-8522, Japan

[2]Faculty of Medicine, Asahikawa Medical University, 1-1-1 Midorigaokahigashinijo, Asahikawa 078-8510, Japan

[3]Co-Mobility Society Research Center, Keio University, 7-1 Shinkawasaki, Saiwai-ku, Kawasaki 212-00322, Japan

ABSTRACT

Eco-driving is expected to contribute to the efforts to fight against global warming by reducing carbon dioxide emissions from vehicles. It is the driving technique that enables low fuel consumption mainly by accelerating gently at the start and avoiding frequent accelerating and decelerating. The effectiveness of eco-driving has been assessed based on a single vehicle; however, it remains to be investigated on multiple vehicles. This paper assesses the effects of eco-driving when following other vehicles on open roads under two different instructions (Japanese eco-driving and German eco-driving) to reduce fuel consumption based on the experiment environment by integrating a driving simulator and a traffic simulator.

KEYWORDS

ITS; Traffic flow; Global environmental problems

1 INTRODUCTION

The Kyoto Protocol, adopted in 1997, required Japan to reduce its greenhouse gas emissions over the period from 2008 to 2012 to 6 % below the level in 1990. In addition, in June 2009, Japan announced a mid-long term target to reduce its greenhouse gas emissions by 25 % from the 1990 levels by 2020. The breakdown of carbon dioxide emissions in Japan indicates that the emissions from transportation account for 20 % of the total, and it is mostly from vehicles. In such situations, eco-driving is attracting attention as one of the ways to reduce carbon dioxide emitted by vehicles.

Eco-driving is considered to be effective in improving the fuel efficiency and is being promoted in various ways. One of the representative activities in Japan is "10 tips for fuel-conserving Eco-Driving" promoted by Team Minus 6 % [1]. "10 tips for fuel-conserving Eco Driving" recommends drivers techniques such as a soft starting with gentle acceleration, called "e-start", and driving without excessive accelerating and decelerating. In this paper, such driving techniques are collectively referred to as Japanese eco-driving. On the other hand, in Europe especially in Germany, the driving technique to accelerate quickly to reach the fuel efficient range as fast as possible, which differs from Japanese eco-driving, is being widely implemented as eco-driving [2]. This driving technique is called German eco-driving. The biggest difference between Japanese and German eco-driving is how to accelerate when the vehicle starts moving. This study aims to evaluate quantitatively the difference in the fuel consumption rates during driving between Japanese eco-driving and German eco-driving.

2 POSITION OF THIS STUDY

The effect of Japanese eco-driving in reducing fuel consumption rates has been verified in various actual driving situations, including the comparison of before and after education for transport drivers such as truck drivers [3, 4] and the comparison of before and after instruction of eco-driving for general drivers [5]. However, there have been few studies of comparison performed in the same conditions.

With regard to German eco-driving, Beusen et al. [6] reported that the fuel consumption was improved after a driver education course on eco-driving was given compared to that of before, through obtaining the driving data of 8 subjects.

As a comparative study on the effects of Japanese eco-driving and German eco-driving, the author et al. [7] analyzed the differences of the effects between Japanese eco-driving and German eco-driving on the fuel consumption rates based on the data of actual vehicle experiments. In this study, the fuel consumption rates were estimated based on the distributions of velocity-acceleration in normal driving, driving with Japanese eco-driving and with German eco-driving. In addition, Kato et al. [8] held an eco-driving test-ride event with actual vehicles to compare the difference between eco-driving while paying attention to gentle acceleration (equivalent to Japanese eco-driving in this paper) and while paying attention to speedier acceleration (equivalent to the German eco-driving in this paper), which proved that both techniques of eco-driving improved the fuel efficiency compared to normal driving.

As a study using a driving simulator (hereafter called "DS"), we can look at the study by Hiraoka et al. [9]. In this study, an experiment was conducted with the condition to drive following the preceding vehicle in JC-08 mode. It concluded that German eco-driving improves fuel efficiency more than Japanese eco-driving does when compared to normal driving. Furthermore, to conduct an empirical study on open roads with traffic lights is cited as a future challenge.

Therefore, this study analyzed while focusing on the vehicle behavior between two intersections with traffic lights, and compared Japanese eco-driving and German eco-driving about the effect to improve fuel efficiency and the driving behaviors over short distances on open roads in DS, which has been rarely discussed in detail until now.

3 EXPERIMENTAL OVERVIEW

3.1 Experiment Environment

In this study, an experiment was performed under an environment with simulated intersections in a virtual space, created by integrating a DS by Keio University and a traffic simulator “AIMSUN” [10] (hereafter called “TS”) (hereafter called “TS/DS integration”). The DS was jointly developed by Keio University and the National Institute for Land and Infrastructure Management [11], and is made by MITSUBISHI PRECISION CO., LTD. The TS is made by TSS-Transport Simulation Systems, S.L.(TSS).

Our study group has performed computer experiments using a TS in the study into the effects of signal control and street parking on smooth traffic flow [12] and in the review and evaluation of public transportation operational procedures [13]. And the study also used a DS in our experiments into the effects of information provision in consideration of human factors of drivers [14, 15]. Experiments with a TS have the advantage that we can evaluate traffic flow through a vehicle group, using a following model of generalized vehicles. On the other hand, experiments with a DS have the advantage that vehicle control data can be quantified based on each driver's driving characteristics. Incorporating these advantages of TS and DS, we created the TS/DS integration to quantify the driving action in a heavy traffic situation or under complicated signal control. This TS/DS integration is for conducting the driving experiment using a DS in the experimental scenario which is developed through use of a TS. In addition, the data output of travel time, delayed time, stopping time and average velocity can be obtained from the TS, and that of acceleration, velocity, engine speed, accelerator pedal input, brake pedal input and the position of the vehicle can be obtained from the DS.

The DS used in this study is composed of eight image screens, front, left and right fronts, left and right sides, left and right rear, and backside (Refer to Figure 1). A driving image from 10 projectors is projected to these screens. A 360-degree field of vision is provided to the driver in the driver seat. The left and right side mirrors and the rearview mirror are the same type as those for actual vehicles, and they capture indirectly the image on the reverse side of the screen. Furthermore, swing equipment with six axes is attached to the vehicle body. All the equipment is collectively controlled using the main calculator.

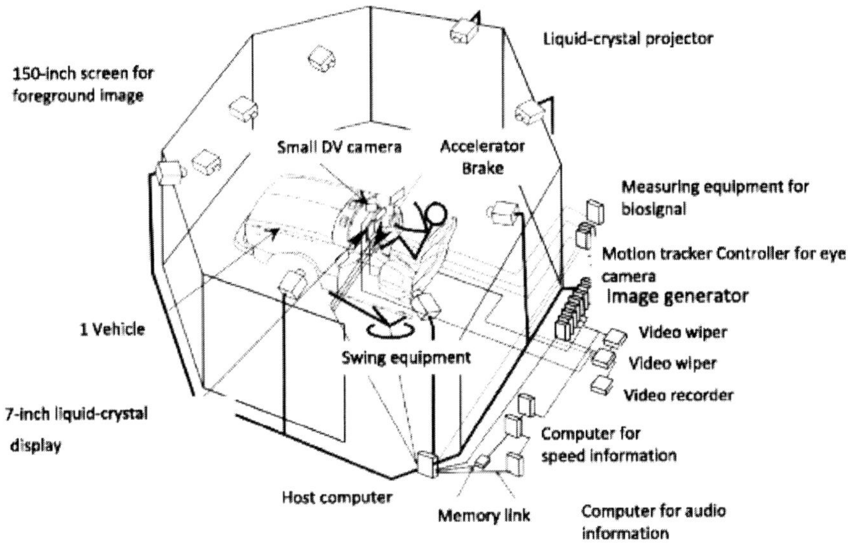

Figure 1. General picture of driving simulator

In this study, an experimental scenario of a straight open road was developed using a TS. The scenario begins with the situation where a vehicle stopped at a traffic light starts moving. It drives straight for 300 m, then must stop at a signalized intersection, and drives straight for 300 m. Each subject got into a DS controlled vehicle which had set that there were vehicles driving in front and at the back. (Refer to Figure 2). The preceding and following vehicles ran according to the vehicle behavior model programmed in the TS [10], and their speed limit was set at 50 km/h. The main TS parameters, which affect drivers of both preceding and following vehicles, were set as in Table 1. As the vehicular swept path of the preceding vehicle, the relation between velocity and distance is shown in Figure 3.

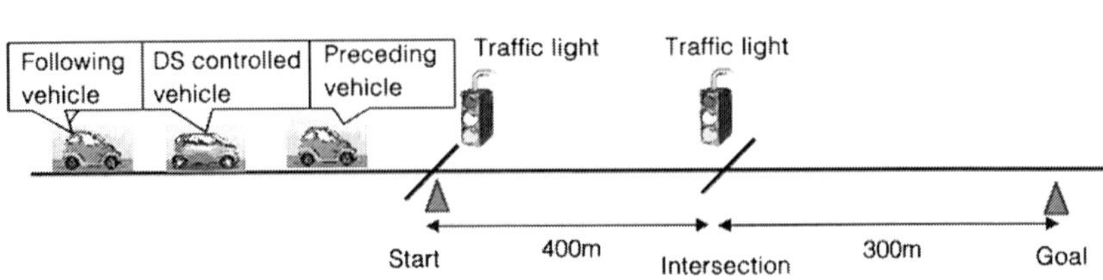

Figure 2. Summary of experiment scenario

Table 1. Main TS parameters

Item (unit)	Mean value (standard deviation)
Maximum desired speed (km/h)	50(0)
Maximum acceleration (m/s^2)	3.0(3.0)
Maximum speed (m/s^2)	6.0(0.5)
Minimum inter-vehicular distance (m)	1.0(0.3)

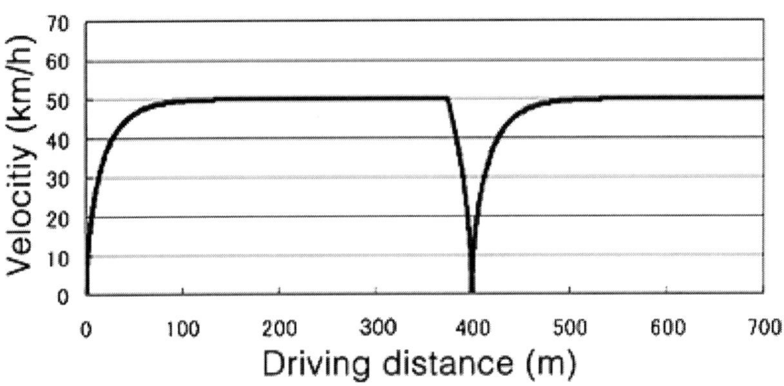

Figure 3. Relationship between velocity and driving distance of preceding vehicle

3.2 Calculation of Fuel Consumption

In this study, the fuel consumption amount calculated based on the map shown in Figure 4, which is in the same manner as the study by Hiraoka et al. [9]. This map is included in the vehicle motion simulation software, CarSim. It shows the instantaneous fuel consumption amount (kg/s) for engine speed and an operation amount on the accelerator of the ordinary motor vehicle with a 2.5-l gasoline engine. We calculate the fuel consumption rate km/l from this obtained amount, the standard density of gasoline 0.76 kg/l and the vehicle velocity km/h.

Figure 4 shows that in the area with very low accelerator pedal input, the fuel consumption amount does not change remarkably even if the engine speed is increasing. When a vehicle is in a coasting state or a braking state, it is usually included in this area. In the area with high accelerator pedal input, the fuel consumption amount increases as the engine speed increases However, it is rare to accelerate with full force, with the exception in some particular circumstances, for example, when merging onto an expressway or driving up a steep slope. The area with the accelerator pedal input about 0.2 to 0.6, called partial throttle area, can be the area where both of the accelerator pedal input and the engine speed become very influential. In particular, an engine speed exceeding 3,600 rpm will tend to sharply increase the fuel consumption amount. In this study, the instantaneous fuel consumption amount (kg/s) is calculated using the accelerator pedal input and the engine speed obtained with a DS.

Though various models are advocated as fuel consumption amount estimation models, the model in this study was chosen because it could provide the driving behavior of drivers and vehicle control data of the DS based on the behavior, enabling the calculation of the fuel consumption amount using the accelerator pedal input and the engine speed (Figure 4).

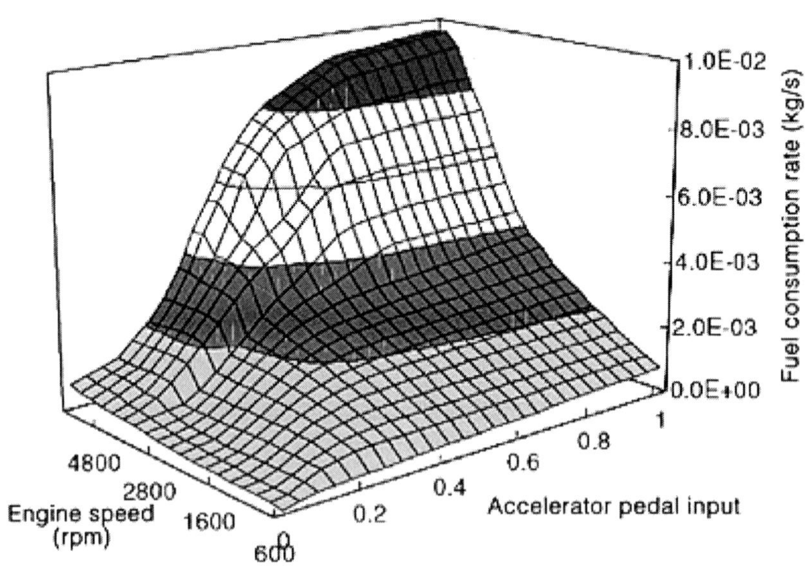

Figure 4. Fuel consumption rate map

3.3 Summary of Attribution of Subjects and Instructions

The subjects are five males (age: 22 to 24 years, average age: 22.6 years, standard deviation: 0.8 years). They filled out the questionnaire about their attribution such as age and experience in eco-driving before the experiment. Table 2 summarizes the attributes of each subject who participated in this experiment.

Table 2. Subject's attribution

ID	Age	Driving experience	Driving frequency
A	23	4 years 10 months	1–2 times per week
B	22	2 years 2 months	Once in every 2 months
C	24	2 years 11 months	1–2 times per month
D	22	2 years 11 months	1–2 times per week
E	22	4 years 1 month	1–2 times per month

The subjects drove a total of 15 times, 5 times each with normal driving, Japanese eco-driving, German eco-driving in this order. The reason for this order is because it was considered that Japanese and German eco-driving would affect normal driving if the test with normal driving was carried out after those with Japanese and/or German eco-driving. As for the order of Japanese and German eco-driving, it is preferable that the order be randomly set for each subject, in consideration of the order effect. In the study, the order was set to avoid the psychological bias in subjects on safety issues such as getting too close to the preceding vehicle if German eco-driving precedes Japanese one. As only 5 subjects participated in the study, the number of driving times was increased so as to obtain more data samples for analysis. As a future challenge, further validation of the experiment with more subjects is necessary. The subjects received instruction on the driving technique and practiced how to drive, before the first test of each driving technique. The instruction is summarized as below.

- Normal driving

  Drive in the same way as they do in their daily lives.

- Japanese eco-driving

  1) Press on the accelerator gently using creep phenomenon to start moving (approximately 20 km/h in 5 s).

  2) Minimize accelerating and decelerating as much as possible and maintain steady acceleration.

  3) Release the accelerator early and actively utilize engine braking.

- German eco-driving

  1) Accelerate as fast as possible (but not too quickly) to reach the fuel efficient velocity range (40 to 80 km/h).

  2) Release the accelerator after reaching the fuel efficient velocity range and utilize coasting actively.

4 EXPERIMENTAL RESULTS

4.1 Comparison of Travel Time

In this experiment, the travel time from the starting point to the goal in experimental scenario of Figure 3 was compared between the normal driving, Japanese eco-driving and German eco-driving. The result is shown in Figure 5.

When the result went through one-way analysis of variance, the travel time with Japanese eco-driving is longest of all and is longer than that with normal driving by 1.8 % (1.4 s), and with German eco-driving by 2.1 % (1.8 s) ($p < 0.1$), as shown in Figure 5. Why the travel time with Japanese eco-driving is the longest of all is considered to be due to "e-start", one of the characteristics of Japanese eco-driving, which has gentle acceleration. On the other hand, there is no statistically significant difference in the travel time between German eco-driving which accelerates as fast as possible and normal driving. Based on these findings, the comparisons of German eco-driving and Japanese eco-driving with regard to the way of accelerating reveal that "e-start" used in Japanese eco-driving affects travel time more.

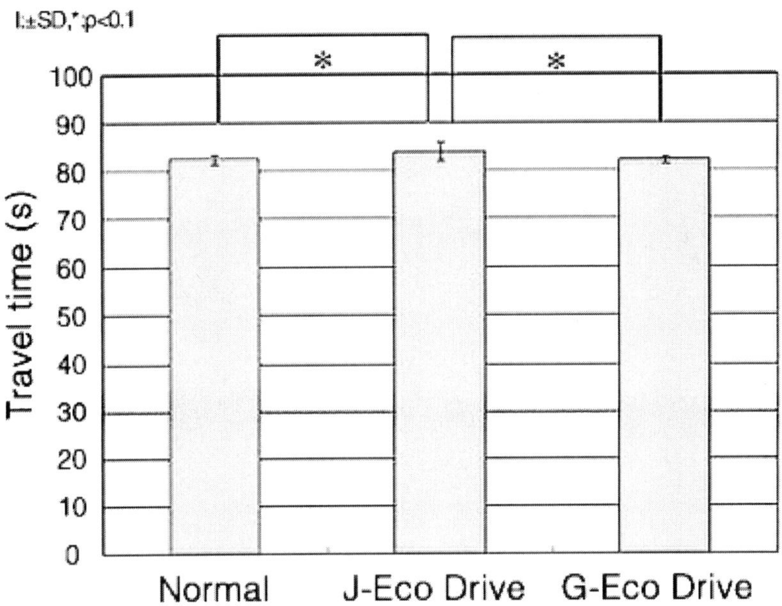

Figure 5. Comparison of travel time ($n = 25$)

4.2 Comparison of Fuel Consumption Rate

In this chapter, the fuel consumption rates are compared among each of the driving techniques. They are calculated using the fuel consumption rate map, shown in Figure 4, based on the accelerator pedal input and the engine speed obtained from the DS. Firstly, the results of the fuel consumption rates for each driving technique are shown in Figure 6.

When the result went through one-way analysis of variance, the fuel consumption rate with German eco-driving was 4.1 % (0.48 km/l) higher than that with normal driving, and that of Japanese eco-driving was 0.87 % (0.1 km/l) lower than that of normal driving, as shown in Figure 6. However, no statistically significant differences were observed in these results.

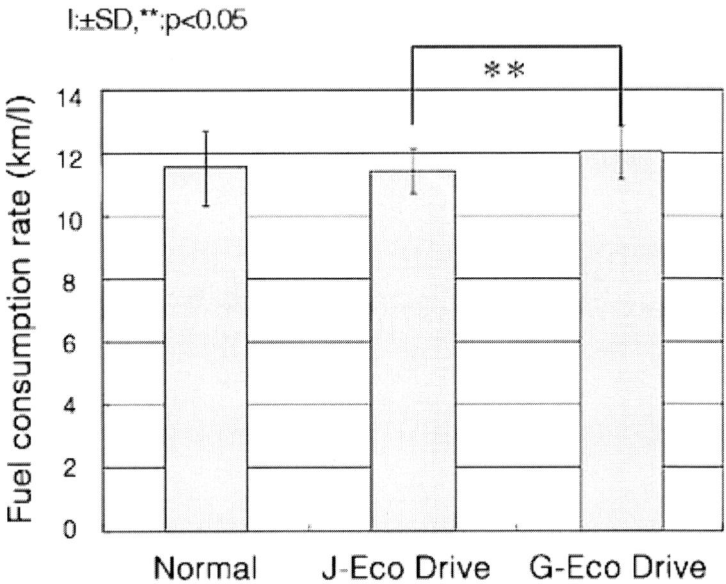

Figure 6. Comparison of fuel consumption rate of each driving technique ($n = 25$)

It is shown that the fuel consumption rate with German eco-driving is 5.1 % (0.56 km/l) higher than that of Japanese eco-driving ($p < 0.05$).

In general, it is said that driving with Japanese eco-driving would reduce the fuel consumption rate compared to normal driving. However, the results of this experiment do not indicate that Japanese eco-driving, which minimizes accelerating and decelerating, will reduce the fuel consumption rate when the intervals between intersections are short. Next, we define the state with 1) no accelerator pedal input nor brake pedal input and 2) a negative acceleration as the coasting state. The ratio of coasting state, which is the proportion of the driving time in the coasting state to the total travel time, is calculated. The result is shown in Figure 7. Figure 7 shows that the ratio of coasting state with German eco-driving is by far the highest, and that with Japanese eco-driving is the lowest ($p < 0.01$).

Additionally, as an example of driving distance and accelerator pedal input for each of normal driving, Japanese eco-driving and German eco-driving in the scenario for this experiment, the driving data of the third test by Subject A is shown in Figs. 8, 9 and 10.

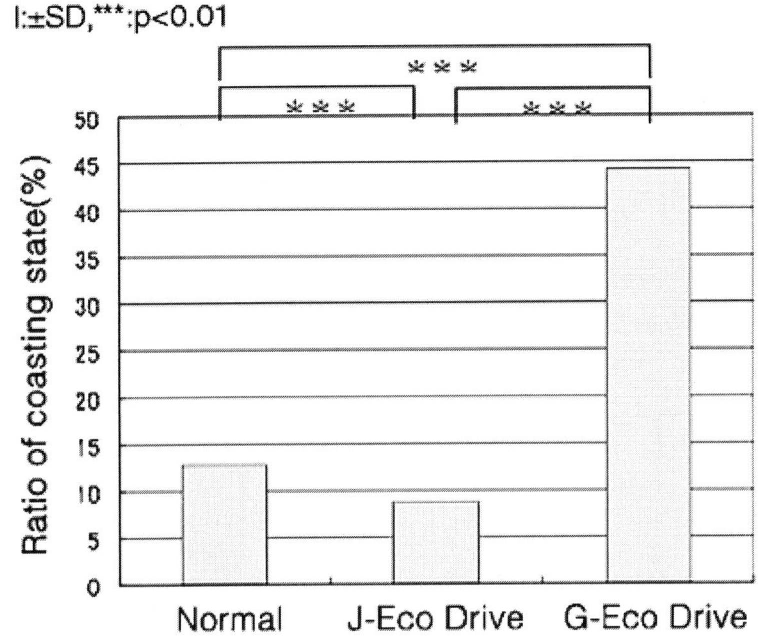

Figure 7. Comparison of ratio of coasting state of each driving technique ($n = 25$)

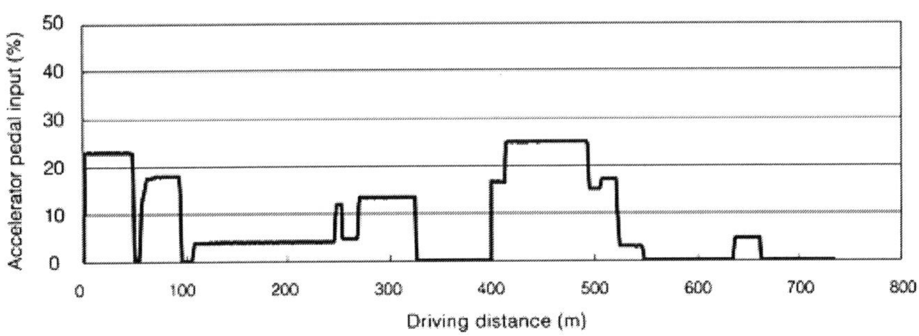

Figure 8. Example of accelerator pedal input and driving distance (Normal driving)

As shown in Figure 9, with Japanese eco-driving, the duration the accelerator is pressed is long even though the accelerator pedal input is small because Japanese eco-driving utilizes "e-start", which made its ratio in the coasting state small. This can be the reason why the fuel consumption rate with Japanese eco-driving was worse than that with German eco-driving.

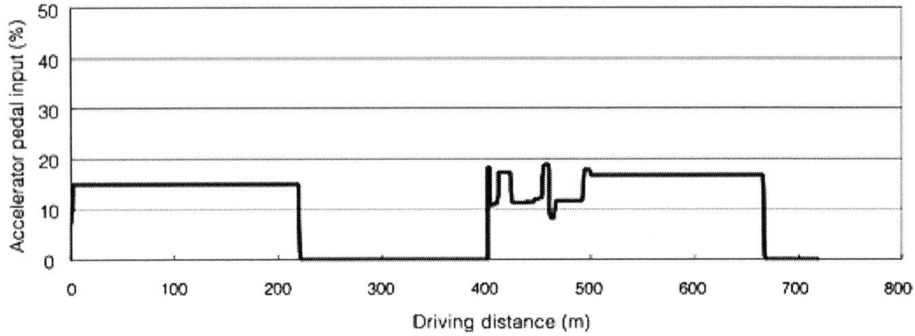

Figure 9. Example of accelerator pedal input and driving distance (Japanese eco-driving)

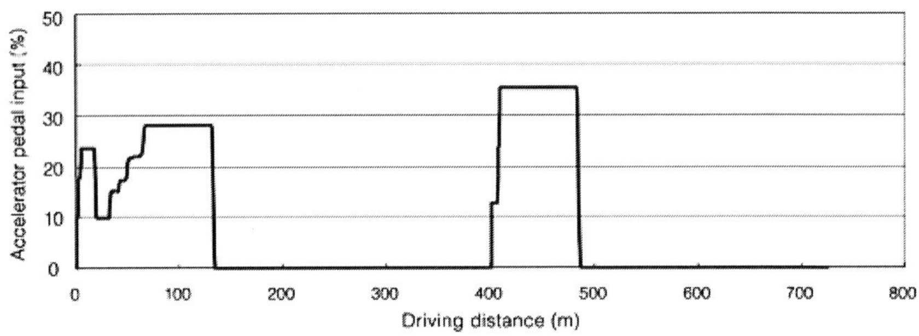

Figure 10. Example of accelerator pedal input and driving distance (German eco-driving)

On the other hand, as shown in Figure 10, the fuel consumption rate with German eco-driving right after the start is higher because the accelerator pedal input is high at the start. It is considered that its fuel consumption rate is better than those of the other driving techniques because when it goes into the coasting state, it requires little fuel consumption after that.

Thus, this study suggested that Japanese eco-driving would not always have an effect to reduce the fuel consumption rate compared to normal driving if a vehicle starts moving and stops within a short distance, although many previous studies have reported that as to the effectiveness of eco-driving, Japanese eco-driving improves the fuel consumption rate.

In addition, we quantified the fuel consumption rates in driving a short distance with Japanese eco-driving and German eco-driving, which was regarded as a problem in the study by Hiraoka et al. [9]. As a result, we could suggest the possibility to improve fuel consumption rates with German eco-driving in such a case where the driving distance from starting to stopping is short. Further validation should be necessary in the future, on the influence of differences in traffic situations such as distance between traffic signals, traffic volume, mix rate of large-sized vehicles and so on.

4.3 Comparison of Driving Techniques and Headway Distance

Figure 11 shows the result of one-way analysis of variance on headway distances with the preceding vehicle for normal driving, Japanese and German eco-driving. Comparison of each driving technique based on this Figure shows the trend that Japanese eco-driving created longer headway distance than normal driving and German eco-driving.

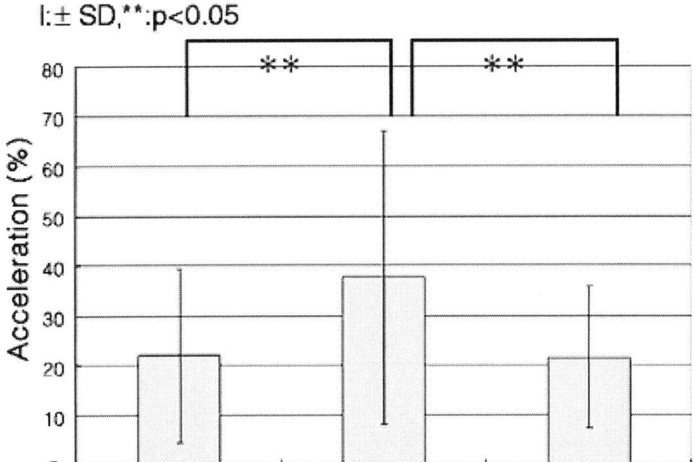

Figure 11. Headway distance of each driving technique

This is also considered to be caused by the difference between Japanese eco-driving with gentle acceleration and German eco-driving with acceleration as quickly as possible to the fuel efficient velocity range.

As shown in Figure 11, comparison of each driving technique with regard to the headway distance indicates that the distance with Japanese eco-driving is twice as long as those of normal driving and German eco-driving. In addition, a comparison of each driving technique shows that the headway distance with Japanese eco-driving varies far more widely than those with the others.

5 CONCLUSION

In this study, the fuel consumption rates with normal driving, Japanese eco-driving and German eco-driving on a straight open road were compared quantitatively using TS/DS integration, which enables the performance of experiments repeatedly under the same conditions. Though the experiment was only performed on a 700-m open road, the result indicated that German eco-driving reduced the fuel consumption rate by approximately 4 % compared to Japanese eco-driving.

In addition, there were no statistically significant differences between normal driving and Japanese eco-driving or German eco-driving. This result suggests that Japanese eco-driving would not be always helpful to reduce the fuel consumption if the distance from the start to stopping is shortened due to a traffic signal, or other reasons.

Furthermore, the relationship between driving manner and headway distance were shown, focusing on the difference in the driving behavior for each style of eco-driving. The result suggested that Japanese eco-driving creates headway distance about twice as long as that by normal driving. It should be noted that this result is not exactly common, as these figures were obtained under the condition that the vehicles should stop two times at traffic signals during their 700-m drive.

In the future, we intend to study quantitatively the effect of eco-driving on the whole traffic flow in the case of high volume traffic. In addition, we intend to study in detail the difference between those subjects who drive with eco-driving techniques routinely and those who don't, by performing an experiment with an increased number of subjects. Moreover, we plan to analyze the changes in fuel consumption amount depending on traffic situations, distance between traffic signals, instructions on eco-driving and so on. Furthermore, we plan to try a new condition of experiment where a DS controlled vehicle is driven while reading the movement of the preceding vehicle in order to drive in the most fuel efficient manner.

REFERENCES

1. The Eco-Drive Promotion Liaison Committee: 10 tips for fuel-conserving Eco Driving. (2005)

2. Ford-Werke GmbH: Ford Eco-Driving Kompakt, http://www.ford-eco-driving.de/

3. Maji, H., Kasuga, N., Ishi, T., Daisho, Y.: Effects of eco-drive activity on improvement in fuel economy and reduction in traffic accidents. Proc. JSAE **38**(3), 99–104 (2007)

4. Saito, A., Takada, H.: The relationship between eco-driving patterns of light duty freight vehicles and the improvement of fuel economy in real traffic conditions. Proceedings of JSAE before lecture presentation, No.99–08. pp. 27–30 (2008)

5. Miyasaka, T., Taniguchi, M., Sanbuichi, H.:Potentiality of effects by driving operation for fuel conservation to automatic transmission car. Proceedings of JSAE before lecture presentation, No.80–05. pp.13–16 (2005)

6. Beusen, B., Denys, T.: Experiences using an onboard logging device with CAN-interface, GPS, GPRS, and Wi-Fi for surveying long term driving behavior. Proceedings of 15th World Congress on Intelligent Transport Systems, CD–ROM. (2008)

7. Matsumoto, S., Kobayashi, I., Nakamura, A., Kawashima, H.: Economic evaluation of influence on traffic flow by eco-driving. Annual report of JSTE, Japan Society of Transportation Economics. pp. 169–178 (2007)

8. Kato, H., Kobayashi, S.: Factors contributing to improved fuel economy in eco-drive. J. JSAE **62**(11), 79–84 (2008

9. Hiraoka, T., Terakado, Y., Matsumoto, S., Yamabe, S.:Quantitative evaluation of eco-driving on fuel consumption based on driving

simulator experiments. Proceedings of 7th World Congress on Intelligent Transport Systems. pp. 163–168 (2008)

10. Babazono, K.: Introduction of microsimulation environment tool, AIMSUN NG. IPSJ SIG Technical Reports ITS **2005**(61), 13–20 (2005)

11. Munehiro, Y., Yamazaki I., Daimon T., Arizumi M.: A counter-measure for crossing collisions based on analysis of human error using driving simulator. Infrastruct. Plan. Manag. JSCE/Proc. **34** (2006), CD–ROM

12. Nakamura, A., Matsumoto, S., Todoroki, T.: Measuring external costs of on-street parking near the bus-stop: the case of Chiba. Annual report of JSTE, Japan Society of Transportation Economics, pp.159–168 (2008)

13. Matsumoto, S., Kumagai, Y., Kawashima, H.: A study on the traffic signal strategy in city traffic in consideration of streetcar. Infrastruct. Plan. Manag. JSCE/Proc **21**(3), 677–685 (2004)

14. Matsumoto, S., Tozawa, T., Hiraoka, T., Yamabe, S., Kawashima, H.: A fundamental study on the effect of eco-driving on following vehicles. Infrastruct. Plan. Manag. Proc. D3 **67**(5), I_1023–I_1028 (2012)

15. Hiraoka, T., Terakado, Y., Matsumoto, S., Yamabe, S.: Effect of eco-driving instruction and fuel consumption meter on fuel consumption reduction. Hum. Interf. Soc. J. **12**(1), 71–78 (2011)

Chapter 9

Social Internet of Vehicles for Smart Cities

Leandros A. Maglaras *, Ali H. Al-Bayatti, Ying He, Isabel Wagner and Helge Janicke

School of Computer Science and Informatics, De Montfort University, The Gateway, Leicester LE1 9BH, UK

ABSTRACT

Digital devices are becoming increasingly ubiquitous and interconnected. Their evolution to intelligent parts of a digital ecosystem creates novel applications with so far unresolved security issues. A particular example is a vehicle. As vehicles evolve from simple means of transportation to smart entities with new sensing and communication capabilities, they become active members of a smart city. The Internet of Vehicles (IoV) consists of vehicles that communicate with each other and with public networks through V2V (vehicle-to-vehicle), V2I (vehicle-to-infrastructure) and V2P (vehicle-to-pedestrian) interactions, which enables both the collection and the real-time sharing of critical information about the condition on the road network. The Social Internet of Things (SIoT) introduces social relationships among objects, creating a social network where the participants are not humans, but intelligent objects. In this article, we explore the concept of the Social Internet of Vehicles (SIoV), a network that enables social interactions both among vehicles and among drivers. We discuss technologies and components of the SIoV, possible applications and issues of security, privacy and trust that are likely to arise.

KEYWORDS

Social Internet of Things; Internet of Vehicles; security; privacy; trust

1. INTRODUCTION

Smart cities are areas where innovation is supported through digital networks and applications [1]. Smart cities are often called sustainable, digital or connected cities [2]. The goal of converting a city into a smart environment is to alleviate the problems resulting from urbanization and increased urban population. A smart city is an urban area that provides the conditions for sustainable economic growth and quality of life. Smart solutions, like traffic congestion avoidance [3], green buildings [4] and modern industrial control systems (ICS) [5], are some of the technologies that can make today's urbanization sustainable. A smart city involves the intelligent use of technology to improve how people live, work, commute and share information [6]. A key aspect of a smart city is next generation vehicles that incorporate new sensing, communication and social capabilities as part of the wider Internet of Things concept. By providing mobile wireless sensing and communications, vehicles can facilitate data access, which is fundamental to make smart cities a reality.

Wireless networks can be divided into three main categories. First, infrastructure wireless networks rely on a central station that coordinates all communications [7]. Second, non-structured networks or *ad hoc* networks give equal roles to all stations in the network [8,9]. Third, hybrid networks combine the first two categories [10]. An example of hybrid networks is hybrid vehicular *ad hoc* networks (hybrid-VANETs) [11], which use *ad hoc* networks for communication between vehicles and infrastructure networks, such as wireless local area networks (WLANs) and cellular systems for communication with a core network [12,13]. Smart vehicles, through their advanced communication capabilities, will be able to interact not only with navigation and broadcast satellites, but also with passenger smart phones, roadside units and other smart vehicles, making them an important component of IoT and the development of smart cities [14]. VANETs combine these with new applications and methods to enable the intelligent communication between vehicles and the connection to the Internet. VANETs rely on roadside units (RSU) and on-board units (OBU) to facilitate the connectivity and the intelligence of the smart vehicle. RSUs are communication infrastructure units that are positioned next to roads to connect vehicles to a larger infrastructure or to a core network, such as a metropolitan traffic management system. The OBU is a network device integrated into smart vehicles that supports different wireless networks, such as dedicated short-range communication (DSRC) and WLAN. A VANET has a diverse range of applications, from road safety, through the detection and avoidance of traffic accidents [15], traffic control, through reduction of traffic congestion [16], as well as infotainment, through the improvement of driving comfort [17]. Recently, three of the biggest companies that provide digital content and operating systems for mobile devices, Google, Apple and Microsoft, have announced their actions toward taking over the in-car infotainment system with their operating systems dedicated to vehicles and mobile devices (CarPlay, Android Auto and Windows Mobile) [18].

The Social Internet of Things (SIoT) [19] is a network of intelligent objects that have social interactions. The Social Internet of Vehicles (SIoV) [20,21] is an example of a SIoT where the objects are smart vehicles. The authors in [21] designed analytical models of the subsystems involved in the SIoV interaction process and proposed models that could be useful in order to deploy Social Internet of Vehicles (SIoV)-based safety, efficiency or comfort applications. Although the basic rules are the same for both social networks of humans and social networks of vehicles, there are significant differences in terms of the dynamic nature of the entities, their social interactions, the topology of the network, privacy concerns and security issues that arise. In this article, we will present a comprehensive review of the research and analysis of such systems.

Social Internet of Vehicles (SIoV) describes both the social interactions among vehicles [22] and among drivers [23]. As described in [23], a vehicular social network is created when a driver enters an area where other people with common interests or relevant content exist. Contrary to this, Nitti *et al.* [22] describe a vehicular social network as social interactions among cars, which communicate autonomously to look for services (automaker patches or updates) and exchange information relevant to traffic. Given that vehicles are becoming more and more autonomous [24] and that applications supporting social interactions among drivers and passengers are already being developed [25,26] we strongly believe that SIoV will eventually be a network of both drivers, passengers and cars. This new interconnection among entities on different levels (vehicles, drivers, passengers) creates new capabilities, poses new challenges and exposes the network to new threats.

New applications that are based on the social concept of vehicular networking were recently developed. For example, RoadSpeak [27] is a voice chatting system that allow commuters to dynamically enter vehicular social networks on the fly and exchange messages. NaviTweet [26] incorporates the driver's preferences into the navigator's route calculation using traffic voice tweets. Caravan Track [25] is a similar application that allows drivers to share mobility data among a cluster of cars. This application, supported by Ford, can be used to filter incoming information to suit the needs and desires of the driver. Porsche also unveiled its new concept electric car that has the ability to post updates to social media [28]. As we move to the era of the next generation of vehicles, smart vehicles with sensing, communication and sociability capabilities, more applications and technologies are going to be developed that will materialize the idea of SIoV.

This article provides an comprehensive survey of the Social Internet of Vehicles. We first review components and technologies of the SIoV in Section 2. We then discuss context awareness in Section 3. In Section 4, we present how social network analysis methods can be applied in SIoVs. Next, in Section 5, we review security and trust issues related to SIoVs. Finally, we discuss current challenges and open issues regarding driver privacy in Section 6.

2. SIOV TECHNOLOGIES AND COMPONENTS: NEXT GENERATION OF VEHICLES

The design, development and deployment of vehicular networks is boosted by recent advances in context-aware technology and wireless vehicular communication techniques, such as dedicated short-range communications (DSRC), Long-Term Evolution (LTE), IEEE 802.11p and Worldwide Interoperability for Microwave Access (WiMax) [29]. An increasing number of social network applications is being proposed for vehicular networks, which leads to a shift from traditional vehicular networks toward SIoV. In this section, we briefly introduce the key aspects that enable SIoV in current vehicular networks. Similar to [30], we focus on three main components: (1) next-generation vehicles; (2) vehicle context-awareness; and (3) SIoV context-aware applications.

Vehicular *ad hoc* networks (VANETs) are a kind of mobile *ad hoc* network that has been proposed to enhance traffic safety and provide comfort applications to drivers. The unique features of VANETs include fast-moving vehicles that follow pre-determined paths (*i.e.*, roads) and messages with different priority levels. For example, messages for comfort and infotainment applications have low priority, while messages for traffic safety applications require timely and reliable message delivery [31]. Using the on-board unit, vehicles can communicate among themselves (vehicle-to-vehicle, V2V) and with roadside units (vehicle-to-infrastructure, V2I). This enables several other forms of communication, such as vehicle-to-broadband cloud (V2B), where the vehicle communicates with a monitoring data centre, vehicle-to-human (V2H) to communicate with vulnerable road users, for example pedestrians or bicycles, or vehicle-to-sensor (V2S), where the vehicles communicate with sensors embedded in the environment [31].

Vegni *et al.* [32] introduced next generation vehicles as smart vehicles that represent the convergence of communications, infrastructure, computers and autonomy, resulting in new capabilities, such as sensing, navigation, sociability, context-awareness and communication. Based on these aspects, smart vehicles exhibit five features: self-driving, safety driving, social driving, electric vehicles and mobile applications.

2.1. Self-Driving

The aim of autonomous cars is to drive independently without human interaction and to navigate the roads safely to reduce traffic accidents. According to a report by KPMG [33], the use of autonomous vehicles could eliminate 90% of all vehicle accidents. The industry [34] defined six accreditation levels of autonomous driving: At Level 0 (no automation), the driver has control of all aspects of the driving, even if he or she is assisted by various intervention or warning systems. At Level 1 (driver assistance), the vehicle will be fully controlled by the driver. This might include specific warning applications, such as blind spot detection. Level 2 (partial automation) has a low level of automation, including lane detection and an assisted braking system. Level 3 (conditional automation) combines two functions for congested traffic, namely driving in a straight line and automatic speed reduction. Level 4 (high automation) allows limited self-driving, which requires the driver to maintain consistent observation of the roadway in selected environments, to cede control when required. At Level 5 (full automation), the vehicle is capable of reaching the navigated destination without human interaction.

Current mass-market technology offers features on accreditation Level 2 (partial automation). For example, the 2014 Mercedes S-Class comes with options for auto-parking, clever steering, lane keeping, automated acceleration/braking and fatigue monitoring [35]. In traffic jams, the 2014 BMW i3 can accelerate autonomously and brake up to 30 miles per hour [36]. The 2015 Audi A7 provides autonomous braking in traffic jams [37].

In the future, most technology companies and automobile manufacturers predict to offer features on higher accreditation levels. For example, Tesla expects to build a vehicle that can drive 90% of distances in fully-automated mode in 2016. In 2018, Google expects to provide fully-autonomous vehicles on the market. In 2020, Mercedes-Benz, GM, Nissan, Volvo, BMW and Audi all expect to sell fully-automated/self-driving vehicles.

2.2. Safety Driving

The aim of the safety driving feature is to reduce casualties, injures and create vision zero [38], *i.e.*, to reach the ambition of zero traffic accidents. To help achieve this goal, vehicular networks enable applications that provide a wide range of information to drivers and travellers. An example application is described in [39], where the authors propose the use of V2I communications in order to reduce the emergency services' arrival time.

2.3. Social Driving

In social driving, vehicles gather *on-the-fly*to be part of a social network that consists of neighbouring vehicles. These vehicles can share the same interests, locations, move towards the same direction/destination or be part of existing relationships. There exist smart tools that use real-time traffic information, collected by a participatory sensing approach in order to help drivers choose the best route, e.g., Waze [40], Google Live Traffic to Google Maps [41], *etc*. Uber and Lyft, two applications that were recently launched, although having raised many legal issues, also give the users the opportunity to share their cars with people that have similar destinations [42] and can be considered as different expressions of the social driving concept.

2.4. Electric Vehicles

There is high customer growth and interest in fully-electric, hybrid (gas-electric) and zero-emission vehicles. Especially in highly populated areas and low speed zones, these vehicles can contribute to the protection of the environment. Novel methods that deal with the optimal charging of electric vehicles, in static or dynamic mode, have recently been introduced. These vary from stationary stations that are scattered across the road network in central positions [43,44,45], dynamic wireless charging methods that take advantage of the mobility of nodes [46,47] and eco-routing algorithms that run in isolation in every vehicle or in a central way for a fleet of vehicles [48,49]. The authors in [50] present a method for dynamic wireless charging of vehicles based on the concept that vehicles meet at rendezvous points, giving a social aspect to the power transfer procedure.

2.5. Mobile Applications

The integration of network adapters, on-board units, different types of sensors and Global Positioning system (GPS) receivers enables vehicles to gather and analyse information about themselves and the surrounding environment and disseminate this information to nearby vehicles [31].

Depending on their primary purpose, VANET applications are classified into comfort applications and safety applications. Comfort or entertainment applications provide ease, high levels of comfort and tranquillity to all passengers, including the driver. This can include information of current restaurant menus, prices and discounts on highway services, nearby gas stations, shop discounts and current prices for nearby hotels. For passengers, this can also include online gaming, Internet access and social instant messaging [51]. Safety applications use dedicated short-range communications (DSRC) to improve road safety and avoid accidents or reduce their severity.

These five features will expose SIoV to a plethora of sensor data that need to be analysed, to enable vehicles to become context-aware.

In their seminal work, Alam *et al.* [20] envisioned the concept of a social network among cars as a cyber-physical layer on top of the physical vehicular network. Based on this concept, the authors proposed application scenarios that target both drivers, authorities and cars and discussed privacy issues that arise from such a system. Using this work as a basis, we elaborate in the following sections on how next generation vehicles can achieve higher levels of vehicle context-awareness. This is followed by a discussion as to how approaches to social network analysis can be applied within an SIoVs and what new security and privacy challenges arise.

3. VEHICLE CONTEXT-AWARENESS

A key aspect in IoV is to enable vehicles to be context-aware, *i.e.*, to be aware of the circumstances that exist around the vehicle, especially those that are particularly relevant to it [52]. Context-aware systems are those that have the capability to adapt their behaviour to their current contextual environment. Context-awareness in vehicles can be provided by three main subsystems: sensing, reasoning and acting [53,54].

The sensing subsystem gathers contextual information from different sensors integrated with the vehicle's OBU. The type of these sensors differs according to the vehicle's requirements, for example location, infrared or ultrasound. In other terms, this phase represents the way context data are collected.

The reasoning subsystem processes raw data to extract high-level contextual information, such as the driver's situation. Contextual information can be either extracted from a single sensor, defining certain contextual information, or extracted from multiple sensors, defining uncertain contextual information. Detecting driver's fatigue levels is considered uncertain (high-level) contextual information.

The acting subsystem represents the application enforcer, which provides services to users or other drivers. Disseminating warning messages, in-vehicle alerts and smart assisted parking are examples of high-level applications deployed to prevent accidents and reduce road congestion.

Several context-aware systems and frameworks for the SIoV have been proposed in the literature. Hu *et al.* [55] presented s-frame, a framework for social vehicular networks formed by in-vehicle or mobile equipment used by passengers, vulnerable road users and drivers that supports high-level context-aware applications.

Alhammad *et al.* [54] designed a VANET on-street context-aware smart parking assisting system, deploying the concept of the centralized InfoStation to locate and reserve a parking slot. This reservation process is dictated by the driver's preferences. All parking zones have a dedicated InfoStation that acts as a terminal, providing wireless coverage over DSRC, as shown in Figure 1.

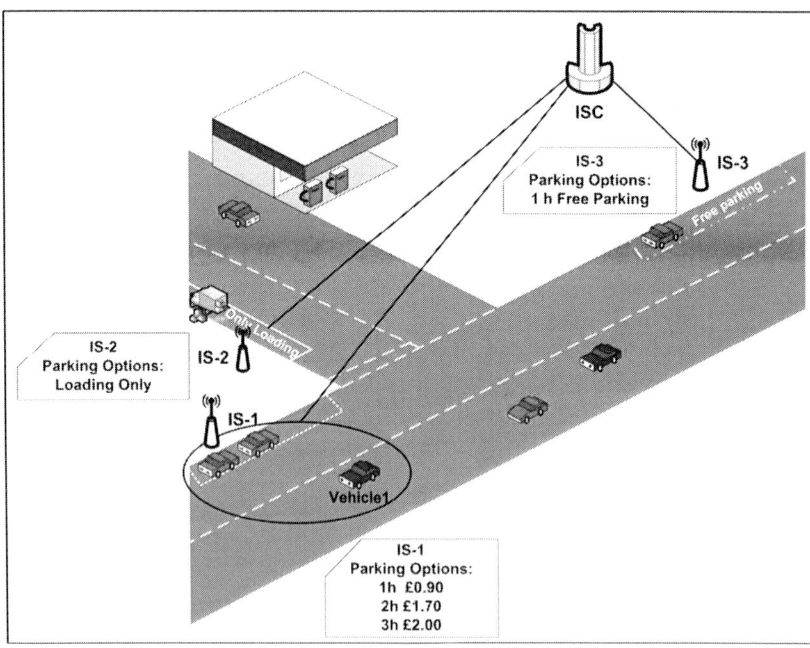

Figure 1. On-street parking system scenario.

Wan *et al.* [29] presented a cloud-based context-aware dynamic parking service that provides planning services for traffic authorities, a parking reservation service and context-aware optimization. Their framework allows drivers to park their vehicle alongside the road for short periods, provided this does not impede the traffic flow. To provide this service, the framework considers contextual information, such as time (e.g., rush hours) and road conditions (e.g., the width of a road). Traffic authorities can manage this service dynamically and effectively reduce parking problems in a smart city.

Shu *et al.* [56] designed SocialDrive to allow drivers to disseminate real-time travel information on social networks and to help them understand their own driving behaviour. SocialDrive provides contextual data to drivers to reflect the ultimate driving style, which can reduce fuel consumption by eliminating unwanted habits.

4. SOCIAL NETWORK ANALYSIS IN SIOVS

Social network analysis (SNA) can be used to discover important players in a network and refers to the use of network theory to analyse social networks. Individual actors within the network, which can either be vehicles, drivers or even passengers, are represented by nodes, and the interactions or relationships among them are represented by edges [57]. Based on the interaction among entities of the network, which can be either static or dynamic, metrics, like centrality, cohesion, degree and clustering coefficient, can reveal relations among nodes, as well as groups of entities that share common habits.

Cunha *et al.* [58] showed that vehicles tend to show a similar behaviour and routines in terms of mobility. The mobility of vehicles can be mapped as a social network, following the same basic laws of degree distribution and distance among nodes. Applying SNA on vehicular networks can therefore improve the performance of communication protocols and services. Traditional concepts from graph theory, like centrality and clustering, can be applied in vehicular networks, as long as they incorporate their specific features, such as mobility of nodes, channel conditions and drivers' behaviour.

The centrality of nodes plays a crucial role in information spreading in a network [59,60]. In a similar way, central nodes can serve as good spreaders of infections [61] or as good points for building defence mechanisms [62]. Furthermore, central nodes can be elected as the clusterhead of groups that are created on the fly. A clusterhead may act as a relay node for traffic coming or destined to different clusters or as a relay node for intra-cluster communication [63].

4.1. Centrality of Nodes in an SIoV

Centrality metrics have been developed and used in the SNA to characterize and measure the importance of individual entities in a social network. The objective of these metrics is to find nodes that are central in a graph and can thus serve as relay nodes in terms of efficient information dissemination. Generally, a node with high centrality is more central compared to its neighbours, although it is highly affected by the network topology and the application that is used.

Betweenness centrality is a good metric for identifying nodes that can participate in the forwarding procedure of data [64], especially when used in static networks. According to betweenness centrality, a node is central to the degree that it stands between others. For example, in Figure 2, nodes *A* and *B* are the most central nodes, because they participate in the majority of shortest paths between the rest of the nodes. Betweenness centrality has been adapted for different types of networks, e.g., sensor networks, where the remaining energy of nodes influences the selection of the next hop in a forwarding chain [65], and delay-tolerant networks, where the delay that every message faces is affected by the popularity of nodes [66].

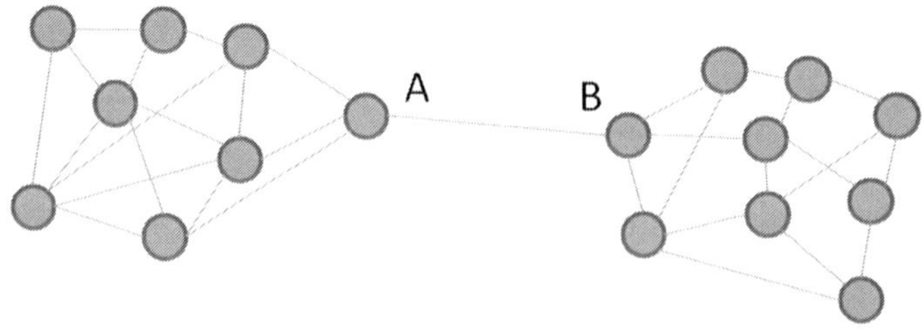

Figure 2. Betweenness centrality.

In mobile *ad hoc* networks, centrality metrics represent the significance of nodes at any time instance. Daly *et al.* [67] show an effective use of social network and centrality analysis in mobile networking. They derive a social routing algorithm from a combination of similarity and betweenness centrality. When the exact location of the destination node is unknown, the algorithm forwards messages to the nearest central node, thereby increasing the potential of selecting a suitable carrier of the information towards the destination.

Aside from betweenness centrality, there are similar metrics that combine graph theory with social characteristics. In degree centrality, central actors are the ones that have the most ties in the network graph. Closeness centrality focuses on how close one actor is to the other. The idea is that an actor is central if it can quickly interact with all others, and it is based on the geodesic distances among nodes. Smart vehicles that are equipped with GPS and communication capabilities can make use of closeness centrality to dynamically identify the central nodes that can be good candidates for forwarding important messages [68].

Maglaras *et al.* [69] proposed to rank vehicles based on the road segments that they are going to follow. They start by ranking each road segment according to how many vehicles that enter the network are going to traverse it, based on their historic data. For vehicles with no previous historic data, the shortest path in terms of distance is assumed as the preferred path. Then, each vehicle is assigned a unique ranking based on the accumulated ranking of the roads that it is going to traverse. This metric represents the significance of the node in terms of dissemination capabilities, and it ranks the importance of a node based on dynamic features, e.g., the traffic of road segments and the route of the vehicle. Following a similar approach, Bradai and Ahmed [70] rank vehicles based on the relative location to their neighbours and their potential of reaching other vehicles. Based on these characteristics and inspired by SNA, they propose a new centrality metric, called dissemination capacity. The proposed ReViVprotocol adds a rebroadcaster selection module to the existing IEEE DSRC and thereby reduces the reported delay, so that real-time streaming applications become feasible.

The role of central nodes can be played by roadside units (RSUs), which are computing devices located on the roadside that provide connectivity support to passing vehicles. RSUs can be of different kinds, for example cellular base stations or wireless access points, and can provide various kinds of communication capabilities to approaching vehicles. The optimal placement of RSUs is similar to determining the central nodes in a VANET, and it has been widely investigated both for urban [71] and highway [72] environments. Rongxing *et al.* [73] propose a novel Social-based PRivacy-preserving packet forwarding (SPRING) protocol for vehicular networks. In SPRING, the optimal locations where the RSUs must be deployed are intersections where many social interactions happen. The RSUs are used as relay nodes to assist cars in packet forwarding. Similar to this work, Huang *et al.* [74] investigate the RSU optimal placement problem, taking into account location privacy parameters. The authors in [75] present a density-based approach for roadside unit deployment in urban scenarios, where RSUs are placed according to the inverse proportion of vehicles densities, in order to reassure seamless connectivity to the Internet.

4.2. Social Clustering of Vehicles

Clustering is a method widely used in all kinds of networks, ranging from mobile *ad hoc* networks [76] to sensor networks [77]. Clustering exists in different forms and can be performed using a wide range of network characteristics, bringing significant benefits to a network. For example, clustering can alleviate the problem of an overwhelming number of broadcasts in dense networks (broadcast storm) [78], increasing the system throughput and improving the bit error rate. It can significantly decrease packet delays and provide better spectrum utilization in time and space. Clusters that partition the network cleverly can also allow data aggregation and increase network longevity. By creating small groups of nodes that share common characteristics, the network appears to be smaller and more stable. To form clusters, vehicles can combine LTE and DSRC communication capabilities, as shown in Figure 3. Cluster heads can perform special operations inside a cluster, such as regulation of channel use, aggregation of data, scheduling and packet routing.

A special characteristic of a VANET environment is diversity in the mobility patterns that vehicles follow, which is based on the road network's local condition, for example reacting to traffic congestion or accidents. This characteristic makes a VANET a very dynamic network where the vehicle density exhibits large variations. Another problem that the communication between moving vehicles has arises when the receiver and the source are moving towards different directions. The problem that is called Doppler shift has attracted much attention, and many solutions have been proposed [79]. The authors in [80] identified the key factors that affect the delivery of warning messages between vehicles, which are the radio propagation model, the density of vehicles and the roadmap. This work showcased how dynamic a VANET environment can be and how small variations in only one feature can heavily affect the performance of the network.

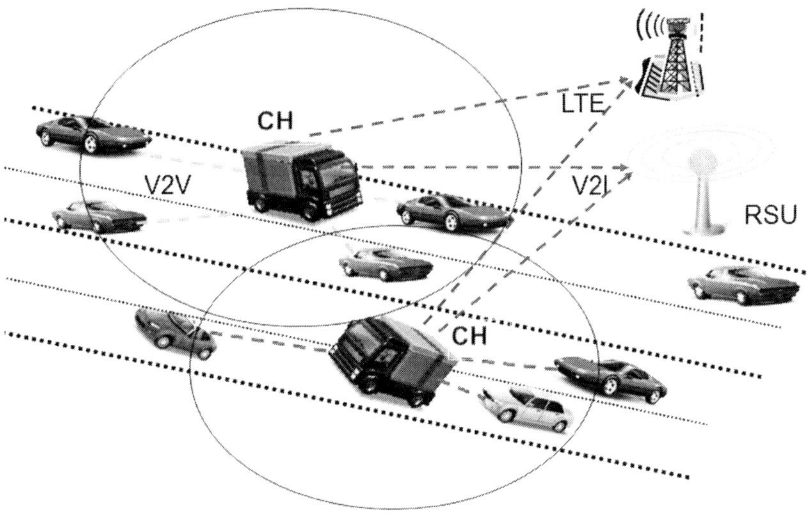

Figure 3. Clustering of vehicles.

Having this dynamic characteristic of a VANET in mind, a good clustering method that can reassure good stability and high cluster lifetime must incorporate many different factors during both the cluster formation and cluster maintenance processes [63,81]. In addition, an effective clustering method has to take into account signal fading and channel interference. Hassanabadi *et al.* [82] presented APROVE, a protocol for distributed election of cluster heads using affinity propagation from a communications perspective. Maglaras *et al.* [81] proposed a distributed clustering algorithm, which applies virtual forces among the nodes to form stable clusters. The applied force is relative to the current and predicted future distance among each pair of vehicles and to their relative velocities. The method assigns positive virtual forces among vehicles that move towards each other and negative forces to vehicles that follow different directions. Nodes decide whether to become cluster heads, join a nearby cluster or leave their cluster depending on their current state and the relation of the total virtual forces applied to it from its neighbours. Vegni *et al.* [83] introduced a new cluster-based routing protocol that reduces network overload, message duplication and packet collisions by allowing vehicles to selectively transmit messages.

When moving on the social aspect of vehicular communications, new parameters, like the frequency of interactions between entities, historical data of driver behaviour and driver habits, must be taken into account. As drivers tend to follow similar routes when moving in a city or on a highway, past mobility information can be used to build distinct social profiles of the drivers. These social profiles can become a basic characteristic to create social groups of vehicles and drivers. In [84], the nodes are divided into different groups based on the regularity of contact between vehicles with fixed routes. Using the social behaviour of vehicles as a basic parameter in the clustering formation procedure, the method presented in [85] manages to increase cluster stability. The social behaviour of drivers is derived from historical data that are collected from RSUs, which are deployed at critical points of the road network. To implement the proposed clustering method, the path that vehicles are likely to follow is added to every beacon message. Both methods use the routes that the vehicles follow or tend to follow as an additional parameter to form stable clusters.

A special category of clusters is platoons. Platooning describes the automated coupling of vehicles by electronic means, with the first vehicle taking control [86]. The lead vehicle decides the speed with which the platoon will move, while the remaining vehicles regulate their speed to follow automatically. The advantages of this scheme are higher security and improved use of resources. Large transport vehicles naturally have a high air resistance, which can be reduced through slipstream effects within a platoon. This saves fuel and helps to make better use of road space [87]. A platoon can consist of an arbitrary number of vehicles, and as vehicles become members of the platoon, they can also form a social network on the fly [88]. After the creation of the platoon, the leading vehicles can serve as the clusterhead and take over most of the communications between the platoon and the outer network (see Figure 4).

5. SIOV SECURITY ISSUES

The security of a vehicular network is a vital aspect of an SIoV, as the compromise of a vehicle can lead to life-threatening situations, as well as a general degradation of other components that use the SIoV as part of a wider smart city infrastructure. Different from a standard computer network, a vehicular network provides a large number of distributed and heterogeneous resources and computational functionalities. The connectivity between vehicles in various geographic locations makes security a more complicated issue to solve, as is the heterogeneous nature in terms of ownership, manufacturer and, indeed, users. Social vehicular network security considers social characteristics and human behaviour [30] alike. This section reviews issues of security, trust and reputation in vehicular networks with a particular focus on the social aspects.

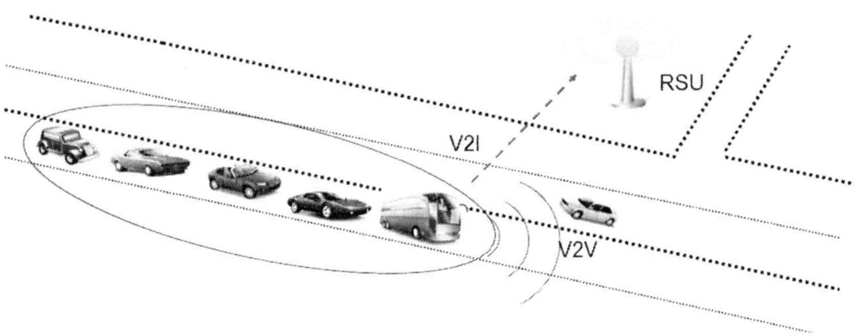

Figure 4. A platoon of vehicles.

5.1. SIoV Threats

Existing research of vehicular security threats considers them as a whole and has not highlighted the social aspects that have a direct impact on the security of the overall infrastructure. Raya reviewed the threats to vehicular networks and classified them into insider/outsider, malicious/rationale and active/passive [89]. Zeadally [90] classified them into threats to availability, threats to authenticity and threats to confidentiality. In the following, we describe the main security threats and highlight some solutions for SIoV security.

5.1.1. Denial of Service Attack

Denial of service (DoS) attacks aim to prevent legitimate users from accessing data or services in computer networks. In vehicular networks, this attack floods and jams the traffic with large volumes of irrelevant messages that negatively impact the communication between the network's nodes, on-board units and roadside units. Because the vehicle under attack is part of the wider infrastructure of the SIoV embedded in a smart city environment, there is a large number of high-powered computing facilities in close proximity to the target. An attacker can potentially use these for jamming attacks against the target's on-board sensory equipment to counter the target vehicle's ability to identify rogue messages through corroboration with its local information sources.

The problem of DoS attacks can be addressed by using a voting scheme [91]. However, voting schemes may fail if the attacker can produce false identities to disguise himself [92].

5.1.2. False Message Injection

An insider attacker can sign a false message and broadcast it to the network. The attacker can thus manipulate the traffic flow and affect the decisions of other drivers, causing damage through traffic jams or accidents.

In vehicular networks, the notion of proof-of-relevance (PoR) has been proposed to address this issue by filtering false data via authentic consensus. PoR is based on the idea that witness vehicles provide authenticated endorsements to information about an event. The event reporter can thus prove that he/she is authentically relevant to the event [93].

5.1.3. Malware

Malware, such as viruses or worms, is usually introduced through outside unit software and firmware updates. Malware can infect vehicles and even allow remote adversaries to take control of individual vehicles. Remote access Trojans, paired with the advanced communication facilities that VANETs bring to the SIoV, can gain control of the internal CAN bus and disrupt essential services. Remote attacks of this form have been widely demonstrated and have shown in dummy tests to risk the safety of drivers and passengers [94]. Other research on malware targeting vehicles has shown that the spread of malware can be achieved through vulnerabilities in the computers used to maintain and diagnose vehicles during servicing.

This has wider implications than just the infection of a single vehicle, because the spread of malicious software is taking place through a trusted service platform, potentially affecting an entire product line. As this is already possible with today's technologies, the potential for malware to spread through the SIoV is even higher, for example when a vehicle's social component receives software or firmware updates from another unknown vehicle.

Existing work proposed the person authentication system [95] to ensure only the authorized user can use the vehicle. Zhang proposed a cloud-assisted vehicle malware defence framework to address the malware challenges [96]. A key challenge is the maintenance of up-to-date patches and signature files. While cloud-assisted frameworks can centralize some of these efforts and make the problem more manageable, the roll-out of automated patch management in a large-scale SIoV is a formidable challenge, given the heterogeneous nature of current vehicles. Other considerations are that the sophistication of advanced persistent threats (APT) is beyond what manufacturers and infrastructure providers are prepared to or can afford to defend against. This means that the defence of individual vehicles will be left to individuals, which based on experience with currently-available ICT infrastructures, is open to a wide array of threats, including masquerading and social engineering attacks on the members of the SIoV.

5.1.4. Masquerading and Sybil

In a masquerading attack, a vehicle fakes its identity and pretends to be legitimate in the vehicle network. Outsiders can conduct attacks, such as injecting false messages. In a Sybil attack, the attacker generates multiple identities and pretends to be multiple legitimate vehicles concurrently. They can artificially disrupt a roadway and affect the decision making of the other drivers through smart routing systems [16]. In this attack mode, a vehicle can claim multiple locations at the same time, which can cause traffic chaos.

Sybil attacks can be detected using resource testing [97], which assumes that vehicles have limited resources. This problem can also be address by using public key cryptography [98], where vehicles are authenticated using public keys. However, these approaches are based on assumptions and have limitations in addressing Sybil attacks. Sybil attacks can be launched on the basis of a large compromised SIoV, because a bot-net that is deployed on an SIoV is difficult to defend against, as the nodes are in a constant flux of reconfiguration. Together with randomized attack patterns, this can lead to denial of service attacks against the wider smart city infrastructure through jamming of low-power sensory infrastructures or the degradation of essential monitoring services.

5.1.5. Impersonation Attack

In an impersonation attack, the attacker steals the identity of a legitimate vehicle and can then broadcast security messages on that vehicle's behalf. These messages can impact other drivers' decision making and create traffic problems.

Chhatwal has proposed an approach called building up secure connection along with key factors (BUCK) to detect and isolate the impersonation attack [99].

5.1.6. Solutions and Future Directions

Tremendous research has been done in protecting vehicular networks from being attacked. These include public key infrastructure [100], trusted architectures, such as the IEEE 1609.2 Security Framework [101], and key management and authentication schemes [102]. However, less research has been conducted in the area of social vehicular network security. With the increasing integration of smart devices and vehicles, human beings will play an important role in V2V (vehicle-to-vehicle), V2R (vehicle-to-road), V2H (vehicle-to-human) and V2S (vehicle-to-sensor) interactions. This calls for new research in two directions. First, research needs to address security issues from a social perspective, such as policy making, law enforcement or security awareness training. Second, research needs to study the interaction between human agents and technical security solutions. Experience can be borrowed from social network security [100], where social aspects are well researched.

5.2. SIoV Trust Issues

Trust has been well studied in social networks. Golbeck introduced the concept of social trust based on sociological foundations, defining trust as "a commitment to an action based on a belief that the future actions of that person will lead to a good outcome" [103]. Golbeck sees social networks as platforms to build mutual trust between entities [104]. However, trust in a social network consisting of vehicles and drivers has not been well studied. This section reviews the trust issues in the SIoV.

5.2.1. Trust Characteristics of SIoV

Trust in the SIoV exhibits five characteristics that are different from trust in traditional social settings.

Uncertainty

Trust in the SIoV is uncertain and dynamic. Because the SIoV is inherently dynamic, information can change rapidly, and trust may only be valid for a short period of time. When establishing trust, the SIoV needs to take into account three uncertainties caused by the SIoV's dynamic nature: the uncertainty of an entity's identity, the uncertainty of an interaction entity's behaviour (intentional or unintentional) and the uncertainty in observations. Existing work in social networks addressed this issue by expressing trust using a continuous variable instead of a discrete-valued variable, as the former can better capture the dynamic feature of the context [105].

Subjectivity

Trust in the SIoV is subjective. Our level of trust depends on how our own actions are affected by the context [106]. Trust decision-making thus exhibits a level of subjective probability that is difficult to predict and monitor. We face an inherent risk in vehicular networks when we make trust decisions within the SIoV. Existing work in social trust addressed this issue by proposing recommendation systems. For example, Yang proposed a recommendation system for online social networks based on Bayesian inference [107]. Users can share content ratings with their friends, allowing them to make informed decisions. This provides an opportunity to choose the option with the smallest level of perceived risk.

Intransitivity

Trust in the SIoV is not always transitive. Trust is transitive when it can be extended outside the scope of the two entities who established it [108]. However, this is not always the case. If A trusts B and B trusts C, this does not necessarily mean that A trusts C. Trust transitivity depends on the extent to which you trust the trustee and the trustee's recommendations. As with subjectivity, research in social networks has developed recommendation systems to make trust-based recommendations by allowing users to share trust ratings with peers.

Context Dependence

Trust in the SIoV is context-dependent. For example, A may trust B as a physician to perform medical checks (in the context of a medical situation), while A would not trust B to perform surgery (in the context of an accident). Similarly, different types of trust need to be established in an SIoV, depending on the given context. In social networks, Jøsang's subjective logic [109] has been widely used to model trust networks. Cerutti uses subjective logic in a decision-making approach that considers the notion of confidence. The approach determines weights associated with trust ratings and outputs a trust degree that depends on the current interaction context [110].

Non-Cooperativeness

Trust in the SIoV is not always cooperative. Trust decisions are usually made cooperatively between different entities. However, not all entities in an SIoV are cooperative. Entities may refuse to cooperate for selfish or malicious reasons. Existing work in social networks aims to separate selfish entities once they are detected and encourages altruistic behaviours with incentive mechanisms. Cho [111] researched the trade-off between selfish and altruistic behaviours in terms of a node's individual welfare, for example saving energy, and global welfare, for example achieving a goal with adequate service availability.

5.2.2. Trust and Reputation

Reputation management is part of trust management. Although the concepts of trust and reputation are often used interchangeably, they are subtly different. Trust is active and indicates whether an entity believes in the trust quality of a peer. Reputation is passive and indicates an opinion about an entity. A reputation system is further classified as positive reputation and negative reputation. Positive entity behaviour is recorded in positive reputation systems, and negative entity behaviour is recorded in negative reputation systems [105]. The authors in [112] present a cooperative neighbour position verification mechanism based on V2V communications among vehicles; it can mitigate the impact of adversary users by excluding malicious nodes from the packet routing mechanism. The aim of the reputation system in the SIoV is to establish trust values for each entity in the SIoV, providing information for other entities to make trust-based decisions. It provides information on whether a peer is trustworthy or not, encourages peers to participate in a trustworthy way and isolates untrustworthy peers from acting in the process. Experience can be borrowed from reputation and trust management in social networks to support trust-based decision-making in the SIoV [113].

Another aspect of the use of trust and reputation within a SIoV is related to the length of interactions between peers in the network. Most work on trust assumes that peers build and establish trust through a number of interactions and that trust can be verified eventually to influence future interactions or recommendations for that peer. While these assumptions hold in the context of electronic transactions and service provisions, they are more difficult to assess in the highly dynamic, reactive environment of the SIoV. In addition, many reputation-based systems rely on the predictability, rationality and repetition of behaviours. In a targeted cyber-attack scenario, these assumptions do not necessarily hold, especially if a malware's spread affects a large number of vehicles in a platoon.

In sum, trust is a multidimensional concept. Trust-based management and decision making in SIoV faces big challenges. Trust establishment needs to address the trust characteristics of the SIoV. Future work should also focus on how to adapt the trust decision frameworks, models and systems in social networks to address similar issues in the SIoV.

6. SIOV PRIVACY ISSUES

The SIoV enables social interactions among vehicles and drivers and thus incorporates features of both vehicular networks and social networks. Both types of networks raise important privacy issues that need to be considered in the context of the SIoV. In this section, we review these privacy issues, discuss privacy-enhancing technologies for the SIoV and highlight research questions that need to be addressed to make SIoVs viable.

6.1. Privacy in Vehicular Networks

The main privacy issue in vehicular networks is location privacy [114]. Vehicles in a vehicular network broadcast unencrypted messages that contain a vehicle identifier along with the vehicle's location, speed and heading. In many cases, these data can be linked to the driver's identity. Using these data, a system that is ultimately designed to offer safety and comfort applications to drivers can be abused by third parties, such as employers, insurance companies or criminal organizations to track individuals [115]. This tracking can reveal sensitive locations, such as home or work locations, along with the time and duration of each visit [116], effectively allowing one to infer the detailed behavioural profiles of drivers.

It is important to note that the location privacy requirements of participants in vehicular networks can conflict with authentication requirements. Because messages in vehicular networks can contain information about safety-critical events, recipients of messages need to make sure that messages have been sent by a trustworthy source before acting on them [117]. If this is realized by strong authentication of message senders, the vehicular network cannot provide privacy at the same time.

Therefore, privacy-preserving solutions for vehicular networks often rely on pseudonyms. A pseudonym takes the place of a vehicle identifier, and it should not be possible to link pseudonyms to the real identifier. Because it can be desirable to allow authorities to find the real origin of a message, for example when misbehaviour has been detected, cryptographic schemes have been designed to provide privacy against ordinary participants, but allow authorities to revoke privacy [118]. To protect against tracking attacks, pseudonyms need to be exchanged frequently, and it is important to ensure that successive pseudonyms cannot be linked to each other [119]. Methods to achieve this unlinkability of pseudonyms include mix zones [120] and silent periods [121]. Recently, a new method that guarantees anonymity while at the same time spotting and revoking malicious users of a VANET is introduced [122].

6.2. Privacy in Social Networks

Privacy issues in social networks include identity theft, stalking, the possibility to create digital dossiers about participants and inadvertent publication of information that was intended to remain private [123]. This is exacerbated by the fact that many social networks have overly complex privacy policies and do not promote available privacy controls to users [124]. In addition, even when users choose to keep most of their information private, several attacks have shown that it is possible to infer sensitive attributes despite restrictive privacy settings [125,126]. This is made worse by the fact that information about a user is often revealed by the user's friends and family, resulting in a situation of interdependent privacy, *i.e.*, where a user's privacy depends on other people [127].

Privacy towards the provider of a social network can be achieved by encrypting all information that a user posts to the social network. Several approaches have been proposed that add encryption to existing social networks without requiring the consent or participation of the provider and without affecting the usability of the social network [128,129,130].

Another privacy concern in social networks is link privacy, *i.e.*, the privacy of relationships between users. Link privacy aims to hide the type and existence of social relationships. This is important because it is possible to re-identify anonymous users in a social network graph [131] and to infer global relationship information by subverting a limited number of user accounts [132]. Mitigation strategies against attacks on link privacy aim to make attacks more expensive and rely on the network provider to implement them [133].

6.3. Privacy-Enhancing Technologies for the SIoV

Combining privacy issues from social networks and vehicular networks, it is easy to see some of the privacy issues that will affect the SIoV. In particular, we envision three risks. First, the privacy threats against social and vehicular networks will become accessible to a wider range of adversaries. For example, it will be possible to eavesdrop on information about social networks via wireless vehicular transmissions, and it will be possible to find location information via online social networks. Second, the information available today will be augmented in two directions: social network profiles will be combined with ubiquitous geo-tagging, and vehicular network messages will be combined with information about a driver's preferences and behaviour. This will allow the inference of much more detailed driver profiles, which can be exploited for marketing or surveillance. Third, it is unclear how privacy policies can be effectively conveyed to a driver in a vehicle and how privacy controls may be offered to drivers. This may lead to even more inaccessible privacy policies and less accessible privacy controls than with today's social networks [124].

To counter these risks, privacy-enhancing technologies for the SIoV are needed. To date, there are very few approaches in the literature that specifically target SIoVs. Most notably, the proposed SIoV architecture in [20] attempts to address privacy issues. However, their approach has three weaknesses. First, their approach relies on cloud storage, which means that the privacy issues associated with cloud computing apply, for example data privacy both towards the cloud provider and the provider's other customers [134]. Second, the privacy protection proposed in their architecture consists of tagging all messages with a privacy value (public, private and protected). This policy-based protection depends on whether the recipient of a message honours the transmitted privacy value and is ineffective against most adversaries. Third and most importantly, it does not include technical privacy protections, such as cryptographic means.

Future privacy-enhancing technologies for SIoVs should be based on proven privacy-enhancing technologies for social networks and vehicular networks. New privacy-enhancing technologies and those resulting from combinations of existing technologies need to be evaluated thoroughly to make sure that they provide an adequate amount of privacy. Because the SIoV is a combination of social and vehicular networks, evaluations need to use a selection of privacy metrics from both domains [135,136].

7. CONCLUSIONS

In this article, we surveyed the key concepts of the Social Internet of Vehicles (SIoV), a new type of network that enables social interactions between vehicles, drivers and passengers in the Internet of Vehicles. More specifically, we reviewed enabling technologies and key components of the SIoV and presented context-aware SIoV applications that can be deployed in a smart city. Three main components were introduced that include next-generation vehicles, vehicle context-awareness and SIoV context-awareness applications. Context-aware systems allow vehicles to adapt their behaviour as per the contextual information gathered from different subsystems, such as sensing, reasoning and acting. Several context-aware frameworks are available that support high-level context-aware applications design. Future work should enhance the context-aware systems by adapting existing frameworks and incorporating context-aware applications from different vendors.

We also identified the types of interactions that can happen between vehicles, drivers and passengers and discussed how social network analysis methods can be used to improve the operation of an SIoV. Using historical data of drivers, vehicles that are moving in a road network can be ranked in terms of their social behaviour. Ranking of vehicles can be used in order to perform efficient message dissemination by selecting the nodes that follow populated roads, so as to alleviate the broadcast storm problem. In order to cope with the interference caused by flooding of messages, clustering of vehicles can also be used, where the social characteristics of the drivers can help in the creation of more stable and robust clustering formations. Combined communication capabilities along with social behaviour of the vehicles can facilitate eco-routing and dynamic charging of electric vehicles. Established social metrics, like degree or betweenness centrality, finally, can be modified in order to cope with the dynamic environment of a vehicular network and provide efficient tools for facilitating the communication among the nodes. Future work should focus on the development of new social metrics that can be dynamic and also able to rank in an efficient manner different entities that co-exist in an SIoV; smart vehicles, passengers, road users and drivers.

Finally, we discussed the issues of security, trust and privacy that SIoVs faces and highlighted how existing security solutions can be applied to these highly dynamic networks. Current security solutions place an imbalanced focus on technical aspects over social aspects. However, with the integration of smart devices and vehicles, the human aspect has become vital. Future research should address the social perspective of security and the interaction between human agents and technical solutions. Trust has been well studied in social networks, but less work can be found in the SIoV. Trust in the SIoV has demonstrated five different characteristics, which are uncertainty, subjectivity, intransitivity, context dependency and non-cooperativeness. Future work should focus on the adaption of the trust decision frameworks, models and systems in social networks, addressing the trust characteristics of the SIoV. Future privacy-enhancing technology in SIoV should be based on existing work for both social networks and vehicular networks. Privacy metrics from both domains need to be considered in order to thoroughly evaluate the upcoming SIoV privacy technologies.

REFERENCES

1. Caragliu, A.; Del Bo, C.; Nijkamp, P. Smart cities in Europe. *J. Urban Technol.* **2011**, *18*, 65–82.

2. Ishida, T.; Isbister, K. *Digital Cities: Technologies, Experiences, and Future Perspectives*; Number 1765; Springer Science & Business Media: Berlin, Germany, 2000.

3. Xiong, Z.; Sheng, H.; Rong, W.; Cooper, D.E. Intelligent transportation systems for smart cities: A progress review. *Sci. China Inf. Sci.* **2012**, *55*, 2908–2914.

4. Bowerman, B.; Braverman, J.; Taylor, J.; Todosow, H.; von Wimmersperg, U. The vision of a smart city. In Proceedings of the 2nd International Life Extension Technology Workshop, Paris, France, 28 September 2000; Volume 28.

5. Su, K.; Li, J.; Fu, H. Smart city and the applications. In Proceedings of the 2011 International Conference on Electronics, Communications and Control (ICECC), Zhejiang, China, 9–11 September 2011; pp. 1028–1031.

6. Hollands, R.G. Will the real smart city please stand up? Intelligent, progressive or entrepreneurial? *City* **2008**, *12*, 303–320.

7. El-Hoiydi, A.; Decotignie, J.D. WiseMAC: An ultra low power MAC protocol for the downlink of infrastructure wireless sensor networks. In Proceedings of the Ninth International Symposium on Computers and Communications (ISCC 2004), Alexandria, Egypt, 28 June–1 July 2004; Volume 1, pp. 244–251.

8. Perkins, C.E. *Ad Hoc Networking*; Addison-Wesley Professional; Addison-Wesley: Boston, MA, USA, 2008.

9. Caballero-Gil, P.; Caballero-Gil, C.; Molina-Gil, J. How to build vehicular ad-hoc networks on smartphones. *J. Syst. Architect.* **2013**, *59*, 996–1004.

10. Liu, B.; Liu, Z.; Towsley, D. On the capacity of hybrid wireless networks. In Proceedings of the Twenty-Second Annual Joint Conference of the IEEE Computer and Communications on IEEE Societies (INFOCOM

2003), San Francisco, CA, USA, 30 March–3 April 2003; Volume 2, pp. 1543–1552.

11. Wang, M.; Shan, H.; Lu, R.; Zhang, R.; Shen, X.; Bai, F. Real-time path planning based on hybrid-VANET-enhanced transportation system. *IEEE Trans. Veh. Technol.* **2014**, *64*, 1664–1678.

12. Tornell, S.M.; Patra, S.; Calafate, C.T.; Cano, J.C.; Manzoni, P. GRCBox: Extending smartphone connectivity in vehicular networks. *Int. J. Distrib. Sens. Netw.* **2015**, *2015*, 478064.

13. Marquez-Barja, J.M.; Ahmadi, H.; Tornell, S.M.; Calafate, C.; Cano, J.; Manzoni, P.; Da Silva, L. Breaking the vehicular wireless communications barriers: Vertical handover techniques for heterogeneous networks. *IEEE Trans. Veh. Technol.***2014**, *64*, 5878–5890.

14. Gubbi, J.; Buyya, R.; Marusic, S.; Palaniswami, M. Internet of Things (IoT): Avision, architectural elements, and future directions. *Future Gener. Comput. Syst.* **2013**, *29*, 1645–1660.

15. Joerer, S.; Bloessl, B.; Huber, M.; Jamalipour, A.; Dressler, F. Demo: Simulating the impact of communication performance on road traffic safety at intersections. In Proceedings of the 20th Annual International Conference on Mobile Computing and Networking, Maui, HI, USA, 7–11 September 2014; pp. 287–290.

16. Maglaras, L.A.; Basaras, P.; Katsaros, D. Exploiting vehicular communications for reducing CO_2 emissions in urban environments. In Proceedings of the 2013 International Conference on Connected Vehicles and Expo (ICCVE), Las Vegas, NV, USA, 2–6 December 2013; pp. 32–37.

17. Cheng, H.T.; Shan, H.; Zhuang, W. Infotainment and road safety service support in vehicular networking: From a communication perspective. *Mech. Syst. Signal Process.* **2011**, *25*, 2020–2038.

18. Gruyer, D.; Belaroussi, R.; Revilloud, M. Accurate lateral positioning from map data and road marking detection.*Expert Syst. Appl.* **2016**, *43*, 1–8.

19. Atzori, L.; Iera, A.; Morabito, G.; Nitti, M. The Social Internet of Things (SIoT)—When social networks meet the internet of things: Concept,

architecture and network characterization. *Comput. Netw.* **2012**, *56*, 3594–3608.

20. Alam, K.; Saini, M.; El Saddik, A. Toward social internet of vehicles: Concept, architecture, and applications. *IEEE Access* **2015**, *3*, 343–357.

21. Alam, K.M.; Saini, M.; Saddik, A.E. Workload model based dynamic adaptation of social internet of vehicles. *Sensors***2015**, *15*, 23262–23285.

22. Nitti, M.; Girau, R.; Floris, A.; Atzori, L. On adding the social dimension to the internet of vehicles: Friendship and middleware. In Proceedings of the 2014 IEEE International Black Sea Conference on Communications and Networking (BlackSeaCom), Odessa, Ukraine, 27–30 May 2014; pp. 134–138.

23. Luan, T.; Lu, R.; Shen, X.; Bai, F. Social on the road: Enabling secure and efficient social networking on highways. *IEEE Wirel. Commun.* **2015**, *22*, 44–51.

24. Schwarz, C.; Thomas, G.; Nelson, K.; McCrary, M.; Sclarmann, N.; Powell, M. *Towards Autonomous Vehicles*; Technical Report 25-1121-0003-117; Mid-America Transportation Center: Lincoln, NE, USA, 2013.

25. Squatriglia, C. Ford's Tweeting Car Embarks on American Journey 2.0. Wired. 2010. Available online: http://www.wired.com/2010/05/ford-american-journey/ (accessed on 15 January 2016).

26. Sha, W.; Kwak, D.; Nath, B.; Iftode, L. Social vehicle navigation: Integrating shared driving experience into vehicle navigation. In Proceedings of the 14th Workshop on Mobile Computing Systems and Applications, Jekyll Island, GA, USA, 26–27 February 2013.

27. Smaldone, S.; Han, L.; Shankar, P.; Iftode, L. RoadSpeak: Enabling voice chat on roadways using vehicular social networks. In Proceedings of the 1st Workshop on Social Network Systems, Glasgow, UK, 1 April 2008; pp. 43–48.

28. Leggett, T. Porsche Concept Challenges Tesla and Posts to Social Media. Available online: http://www. bbc.co.uk/news/technology-34263592 (accessed on 15 January 2016).

29. Wan, J.; Zhang, D.; Zhao, S.; Yang, L.; Lloret, J. Context-aware vehicular cyber-physical systems with cloud support: Architecture, challenges, and solutions. *IEEE Commun. Mag.* **2014**, *52*, 106–113.

30. Vegni, A.; Loscri, V. A Survey on Vehicular Social Networks. *IEEE Commun. Surv. Tutor.* **2015**, *17*, 2397–2419.

31. Al-Sultan, S.; Al-Doori, M.M.; Al-Bayatti, A.H.; Zedan, H. A comprehensive survey on vehicular ad hoc network. *J. Netw. Comput. Appl.* **2014**, *37*, 380–392.

32. Vegni, A.M.; Biagi, M.; Cusani, R. *Smart Vehicles, Technologies and Main Applications in Vehicular Ad Hoc Networks*; INTECH Open Access Publisher: Rijeka, Croatia, 2013.

33. Silberg, G.; Wallace, R. *Self-Driving Cars: The Next Revolution*; Technical Report; KPMG: Amstelveen, The Netherlands, 2012.

34. SAE On-Road Automated Vehicle Standards Committee. *Taxonomy and Definitions for Terms Related to On-Road Motor Vehicle Automated Driving Systems*; Technical Report J3016_201401; SAE: Hong Kong, China, 2014.

35. Mercedes-Benz UK Ltd. Saloon—Specs, Prices & Offers. Available online: http://www2.mercedesbenz.co.uk/content/unitedkingdom/mpc/mpc_unitedkingdom_website/en/home _mpc/passengercars/ home/new_cars/models/s-class/_w222.flash.html (accessed on 15 January 2016).

36. BMW of North America. BMW I Overview—BMW North America. Available online: http://www. bmwusa.com/bmw/BMWi (accessed on 15 January 2016).

37. Audi of America. Audi Innovation: Performance for the Driver. Available online: http://www.audiusa. com/technology(accessed on 15 January 2016).

38. Ding, Z. The enlightenment of vision zero to China's road safety management. In Proceedings of the 2010 WASE International Conference on Information Engineering (ICIE), Beidaihe, China, 14–15 August 2010; Volume 3, pp. 352–355.

39. Barrachina, J.; Garrido, P.; Fogue, M.; Martinez, F.J.; Cano, J.C.; Calafate, C.T.; Manzoni, P. Reducing emergency services arrival time by using vehicular communications and Evolution Strategies. *Expert Syst. Appl.* **2014**, *41*, 1206–1217.

40. Waze. Real-time maps and traffic information based on the wisdom of the crowd. Available online: http://solsie.com/2009/09/real-time-maps-and-traffic-information-based-on-the-wisdom-of-the-crowd/ (accessed on 15 January 2016).

41. Arrington, M. *Google Redefines GPS Navigation Landscape: Google Maps Navigation for Android 2.0*; TechCrunch: San Francisco, CA, USA, 2009.

42. Portland makes Uber and Lyft legal—For now. Available online: http://www.oregonherald.com/oregon/localnews.cfm?id=8910 (accessed on 15 January 2016).

43. Machiels, N.; Leemput, N.; Geth, F.; van Roy, J.; Buscher, J.; Driesen, J. Design criteria for electric vehicle fast charge infrastructure based on flemish mobility behaviour. *IEEE Trans. Smart Grid* **2014**, *5*, 320–327.

44. Putrus, G.; Suwanapingkarl, P.; Johnston, D.; Bentley, E.; Narayana, M. Impact of electric vehicles on power distribution networks. In Proceedings of the IEEE Vehicle Power and Propulsion Conference (VPPC '09), Dearborn, MI, USA, 7–10 September 2009; pp. 827–831.

45. Ruiz, M.A.; Abdallah, F.A.; Gagnaire, M.; Lascaux, Y. Telewatt: An innovative electric vehicle charging infrastructure over public lighting systems. In Proceedings of the 2013 International Conference on Connected Vehicles and Expo (ICCVE), Las Vegas, NV, USA, 2–6 December 2013; pp. 741–746.

46. Jung, G.; Song, B.; Shin, S.; Lee, S.; Shin, J.; Kim, Y.; Lee, C.; Jung, S. Wireless charging system for On-Line Electric Bus (OLEB) with series-connected road-embedded segment. In Proceedings of the 2013 12th International Conference on Environment and Electrical Engineering (EEEIC), Wroclaw, Poland, 5–8 May 2013; pp. 485–488.

47. Maglaras, L.; Topalis, F.V.; Maglaras, A.L. Cooperative approaches for dymanic wireless charging of Electric Vehicles in a smart city. In

Proceedings of the 2014 IEEE International Energy Conference (ENERGYCON), Cavtat, Croatia, 13–16 May 2014; pp. 1365–1369.

48. Boriboonsomsin, K.; Barth, M.J.; Zhu, W.; Vu, A. Eco-routing navigation system based on multisource historical and real-time traffic information. *IEEE Trans. Intell. Transp. Syst.* **2012**, *13*, 1694–1704.

49. Liu, Z.; Zhang, W.; Ji, X.; Li, K. Optimal planning of charging station for electric vehicle based on particle swarm optimization. In Proceedings of the 2012 IEEE Innovative Smart Grid Technologies—Asia (ISGT Asia), Tianjin, China, 21–24 May 2012; pp. 1–5.

50. Dutta, P. Coordinating rendezvous points for inductive power transfer between electric vehicles to increase effective driving distance. In Proceedings of the 2013 International Conference on Connected Vehicles and Expo (ICCVE), Las Vegas, NV, USA, 2–6 December 2013; pp. 649–653.

51. Jakubiak, J.; Koucheryavy, Y. State of the art and research challenges for VANETs. In Proceedings of the 2008 5th IEEE Consumer Communications and Networking Conference, Las Vegas, NV, USA, 10–12 January 2008.

52. Eskandarian, A. *Handbook of Intelligent Vehicles*; Springer: London, UK, 2012.

53. Wang, Z.; Lu, J. The design and implementation of client about vehicle context-aware system based on Ubiquitous Network. In Proceedings of the 2013 IEEE 4th International Conference on Electronics Information and Emergency Communication (ICEIEC), Beijing, China, 15–17 November 2013; pp. 325–328.

54. Alhammad, A.; Siewe, F.; Al-Bayatti, A. An InfoStation-based context-aware on-street parking system. In Proceedings of the 2012 International Conference on Computer Systems and Industrial Informatics (ICCSII), Sharjah, UAE, 18–20 December 2012; pp. 1–6.

55. Hu, X.; Zhao, J.; Seet, B.C.; Leung, V.; Chu, T.; Chan, H. S-Aframe: Agent-Based Multilayer Framework With Context-Aware Semantic Service for Vehicular Social Networks. *IEEE Trans. Emerg. Top. Comput.* **2015**, *3*, 44–63.

56. Shu, W.; Zhang, G.; Wu, M.Y.; Lu, J.L. A social-network-enabled green transportation system. In Proceedings of the 2013 International Conference on Connected Vehicles and Expo (ICCVE), Las Vegas, NV, USA, 2–6 December 2013; pp. 425–430.

57. Scott, J. *Social Network Analysis*; Sage Publications Ltd: Thousand Oaks, CA, USA, 2012.

58. Cunha, F.; Carneiro Vianna, A.; Mini, R.; Loureiro, A. How effective is to look at a vehicular network under a social perception? In Proceedings of the 2013 IEEE 9th International Conference on Wireless and Mobile Computing, Networking and Communications (WiMob), Lyon, France, 7–9 October 2013; pp. 154–159.

59. Basaras, P.; Katsaros, D.; Tassiulas, L. Detecting influential spreaders in complex, dynamic networks. *Computer* **2013**,*46*, 24–29.

60. Borge-Holthoefer, J.; Rivero, A.; Moreno, Y. Locating privileged spreaders on an online social network. *Phys. Rev. E***2011**, *85*.

61. Canright, G.S.; Engø-Monsen, K. Spreading on networks: A topographic view. *Complexus* **2006**, *3*, 131–146.

62. Noel, S.; Jajodia, S. Optimal IDS sensor placement and alert prioritization using attack graphs. *J. Netw. Syst. Manag.***2008**, *16*, 259–275.

63. Souza, E.; Nikolaidis, I.; Gburzynski, P. A new aggregate local mobility (ALM) clustering algorithm for VANETs. In Proceedings of the 2010 IEEE International Conference on Communications (ICC), Cape Town, South Africa, 23–27 May 2010; pp. 1–5.

64. Bavelas, A. A mathematical model for group structures. *Hum. Organ.* **1948**, *7*, 16–30.

65. Cuzzocrea, A.; Papadimitriou, A.; Katsaros, D.; Manolopoulos, Y. Edge betweenness centrality: A novel algorithm for QoS-based topology control over wireless sensor networks. *J. Netw. Comput. Appl.* **2012**, *35*, 1210–1217.

66. Bali, R.S.; Kumar, N.; Rodrigues, J.J. Clustering in vehicular ad hoc networks: Taxonomy, challenges and solutions.*Veh. Commun.* **2014**, *1*, 134–152.

67. Daly, E.M.; Haahr, M. Social network analysis for routing in disconnected delay-tolerant manets. In Proceedings of the 8th ACM International Symposium on Mobile ad Hoc Networking and Computing, Montreal, QC, Canada, 9–14 September 2007; pp. 32–40.

68. Loulloudes, N.; Pallis, G.; Dikaiakos, M.D. *The Dynamics of Vehicular Networks in Urban Environments*; arXiv preprint arXiv:1007.4106; Cornell University: Ithaca, NY, USA, 2010.

69. Maglaras, L.A.; Stathakidis, E.; Jiang, J. Social aspect of vehicular communications. *EAI Trans. Cloud Syst.* **2015**, *1*, 1–10.

70. Bradai, A.; Ahmed, T. ReViV: Selective rebroadcast mechanism for video streaming over VANET. In Proceedings of the 2014 IEEE 79th Vehicular Technology Conference (VTC Spring), Seoul, Korea, 18–21 May 2014; pp. 1–6.

71. Aslam, B.; Amjad, F.; Zou, C.C. Optimal roadside units placement in urban areas for vehicular networks. In Proceedings of the 2012 IEEE Symposium on Computers and Communications (ISCC), Cappadocia, Turkey, 1–4 July 2012; pp. 000423–000429.

72. Rashidi, M.; Batros, I.; Madsen, T.K.; Riaz, M.T.; Paulin, T. Placement of Road Side Units for floating car data collection in highway scenario. In Proceedings of the 2012 4th International Congress on Ultra Modern Telecommunications and Control Systems and Workshops (ICUMT), St. Petersburg, Russia, 3–5 October 2012; pp. 114–118.

73. Lu, R.; Lin, X.; Shen, X. SPRING: A social-based privacy-preserving packet forwarding protocol for vehicular delay tolerant networks. In Proceedings of the 2010 IEEE INFOCOM, San Diego, CA, USA, 14–19 March 2010; pp. 1–9.

74. Huang, X.; Kang, J.; Yu, R. Optimal roadside unit placement with location privacy enhancement in vehicular social network. In Proceedings of the 2015 5th International Conference on Information Science and Technology (ICIST), Changsha, China, 24–26 April 2015; pp. 254–259.

75. Barrachina, J.; Garrido, P.; Fogue, M.; Martinez, F.J.; Cano, J.C.; Calafate, C.T.; Manzoni, P. D-RSU: A density-based approach for road side unit deployment in urban scenarios. In Proceedings of the 2012

IEEE Intelligent Vehicles Symposium, Madrid, Spain, 3–7 June 2012; pp. 1–6.

76. Yu, J.Y.; Chong, P.H.J. A survey of clustering schemes for mobile ad hoc networks. *IEEE Commun. Surv. Tutor.* **2005**, *7*, 32–48.

77. Younis, O.; Krunz, M.; Ramasubramanian, S. Node clustering in wireless sensor networks: Recent developments and deployment challenges. *IEEE Netw.* **2006**, *20*, 20–25.

78. Tseng, Y.C.; Ni, S.Y.; Chen, Y.S.; Sheu, J.P. The broadcast storm problem in a mobile ad hoc network. *Wirel. Netw.* **2002**,*8*, 153–167.

79. Nyongesa, F.; Djouani, K.; Olwal, T.; Hamam, Y. Doppler Shift Compensation Schemes in VANETs. *Mob. Inf. Syst.***2015**, *2015*, 438159.

80. Fogue, M.; Garrido, P.; Martinez, F.J.; Cano, J.C.; Calafate, C.T.; Manzoni, P. Identifying the key factors affecting warning message dissemination in VANET real urban scenarios. *Sensors* **2013**, *13*, 5220–5250.

81. Maglaras, L.; Katsaros, D. Distributed clustering in vehicular networks. In Proceedings of the 2012 IEEE 8th International Conference on Wireless and Mobile Computing, Networking and Communications (WiMob), Barcelona, Spain, 8–10 October 2012; pp. 593–599.

82. Hassanabadi, B.; Shea, C.; Zhang, L.; Valaee, S. Clustering in vehicular ad hoc networks using affinity propagation. *Ad Hoc Netw.* **2014**, *13*, 535–548.

83. Vegni, A.M.; Stramacci, A.; Natalizio, E. Opportunistic clusters selection in a reliable enhanced broadcast protocol for vehicular ad hoc networks. In Proceedings of the 2013 10th Annual Conference on Wireless On-demand Network Systems and Services (WONS), Banff, AB, Canada, 18–20 March 2013; pp. 95–97.

84. Shi, J.; Wang, X.; Huang, M. Social-Based Routing for Vehicular Ad Hoc Networks in Fixed-Route Transportation Scenarios. In *Wireless Communications, Networking and Applications*; Springer: Berlin, Germany, 2016; pp. 681–691.

85. Maglaras, L.; Katsaros, D. Social clustering of vehicles based on semi-Markov processes. *IEEE Trans. Veh. Technol.***2015**, *65*, 318–332.

86. Jia, D.; Lu, K.; Wang, J.; Zhang, X.; Shen, X. A Survey on platoon-based vehicular cyber-physical systems. *IEEE Commun. Surv. Tutor.* **2015**, *18*.

87. Van Arem, B.; van Driel, C.J.; Visser, R. The impact of cooperative adaptive cruise control on traffic-flow characteristics. *IEEE Trans. Intell. Transp. Syst.* **2006**, *7*, 429–436.

88. Mohaien, A.; Kune, D.F.; Vasserman, E.Y.; Kim, M.; Kim, Y. Secure encounter-based mobile social networks: Requirements, designs, and tradeoffs. *IEEE Trans. Dependable Secur. Comput.* **2013**, *10*, 380–393.

89. Raya, M.; Hubaux, J.P. The security of vehicular ad hoc networks. In Proceedings of the 3rd ACM Workshop on Security of Ad Hoc and Sensor Networks, Alexandria, VA, USA, 7 November 2005; pp. 11–21.

90. Zeadally, S.; Hunt, R.; Chen, Y.S.; Irwin, A.; Hassan, A. Vehicular ad hoc networks (VANETS): Status, results, and challenges. *Telecommun. Syst.* **2012**, *50*, 217–241.

91. Malla, A.M.; Sahu, R.K. Security attacks with an effective solution for DOS attacks in VANET. *Int. J. Comput. Appl.***2013**, *66*, 45–49.

92. Lesser, V.; Ortiz, C.L., Jr.; Tambe, M. *Distributed Sensor Networks: A Multiagent Perspective*; Springer Science & Business Media: Berlin, Germany, 2012; Volume 9.

93. Cao, Z.; Kong, J.; Gerla, M.; Chen, Z.; Hu, J. Filtering false data via authentic consensus in vehicle ad hoc networks.*Int. J. Auton. Adapt. Commun. Syst.* **2010**, *3*, 217–235.

94. Koscher, K.; Czeskis, A.; Roesner, F.; Patel, S.; Kohno, T.; Checkoway, S.; McCoy, D.; Kantor, B.; Anderson, D.; Shacham, H.; *et al.* Experimental security analysis of a modern automobile. In Proceedings of the 2010 IEEE Symposium on Security and Privacy (SP), Oakland, CA, USA, 16–19 May 2010; pp. 447–462.

95. Krishna, A.S.; Hussain, S.A. Smart vehicle security and defending against collaborative attacks by malware. *Int. J. Embed. Softw. Comput.* **2015**.

96. Zhang, T.; Antunes, H.; Aggarwal, S. Defending connected vehicles against malware: Challenges and a solution framework. *IEEE Internet Things J.* **2014**, *1*, 10–21.

97. Newsome, J.; Shi, E.; Song, D.; Perrig, A. The sybil attack in sensor networks: Analysis & defences. In Proceedings of the 3rd International Symposium on Information Processing in Sensor Networks, Berkeley, CA, USA, 26–27 April 2004; pp. 259–268.

98. Raya, M.; Papadimitratos, P.; Hubaux, J.P. Securing vehicular communications. *IEEE Wirel. Commun.* **2006**, *13*, 8–15.

99. Chhatwal, S.S.; Sharma, M. Detection of impersonation attack in VANETs using BUCK Filter and VANET Content Fragile Watermarking (VCFW). In Proceedings of the 2015 International Conference on Computer Communication and Informatics (ICCCI), Coimbatore, India, 8–10 January 2015; pp. 1–5.

100. Kumar, N.; Iqbal, R.; Misra, S.; Rodrigues, J.J. An intelligent approach for building a secure decentralized public key infrastructure in VANET. *J. Comput. Syst. Sci.* **2015**, *81*, 1042–1058.

101. Schweppe, H.; Weyl, B.; Roudier, Y.; Idrees, M.S.; Gendrullis, T.; Wolf, M.; Serme, G.; de Oliveira, S.A.; Grall, H.; Sudholt, M.; *et al.* Securing car2X applications with effective hardware software codesign for vehicular on-board networks. In Proceedings of the 27th Joint VDI/VW Automotive Security Conference, Berlin, Germany, 11–12 October 2011.

102. Lu, R.; Lin, X.; Liang, X.; Shen, X. A dynamic privacy-preserving key management scheme for location-based services in vanets. *IEEE Trans. Intell. Transp. Syst.* **2012**, *13*, 127–139.

103. Golbeck, J. Computing with trust: Definition, properties, and algorithms. In Proceedings of the 2006 Securecomm and Workshops, Baltimore, MD, USA, 28 August–1 September 2006; pp. 1–7.

104. Golbeck, J. *Computing with Social Trust*; Springer Science & Business Media: Berlin, Germany, 2008.

105. Adams, W.J.; Hadjichristofi, G.C.; Davis, N.J., IV. Calculating a node's reputation in a mobile ad hoc network. In Proceedings of the 24th IEEE International Performance, Computing, and Communications Conference, Phoenix, AZ, USA, 7–9 April 2005; pp. 303–307.

106. Gupta, S. A General Context-dependent Trust Model for Controlling Access to Resources. Ph.D. Thesis, Jadavpur University, Kolkata, India, 2012.

107. Yang, X.; Guo, Y.; Liu, Y. Bayesian-inference-based recommendation in online social networks. *IEEE Trans. Parallel Distrib. Syst.* **2013**, *24*, 642–651.

108. Sherchan, W.; Nepal, S.; Paris, C. A survey of trust in social networks. *ACM Comput. Surv.* **2013**, *45*.

109. Jøsang, A.; Hayward, R.; Pope, S. Trust Network Analysis with Subjective Logic. In Proceedings of the 29th Australasian Computer Science Conference, Hobart, Australia, 16–19 January 2006; Volume 48, pp. 85–94.

110. Cerutti, F.; Toniolo, A.; Oren, N.; Norman, T.J. *Context-Dependent Trust Decisions with Subjective Logic*; arXiv preprint arXiv:1309.4994; Cornell University: Ithaca, NY, USA, 2013.

111. Cho, J.H.; Chen, R. On the tradeoff between altruism and selfishness in MANET trust management. *Ad Hoc Netw.***2013**, *11*, 2217–2234.

112. Fogue, M.; Martinez, F.; Garrido, P.; Fiore, M.; Chiasserini, C.F.; Casetti, C.; Cano, J.; Calafate, C.; Manzoni, P. Securing warning message dissemination in vanets using cooperative neighbour position verification. *IEEE Trans. Veh. Technol.* **2014**, *64*, 2538–2550.

113. Cho, J.H.; Swami, A.; Chen, I.R. A survey on trust management for mobile ad hoc networks. *IEEE Commun. Surv. Tutor.* **2011**, *13*, 562–583.

114. Eckhoff, D.; Sommer, C. Driving for Big Data? Privacy Concerns in Vehicular Networking. *IEEE Secur. Priv.* **2014**, *12*, 77–79.

115. Dötzer, F. Privacy Issues in Vehicular Ad Hoc Networks. In *Privacy Enhancing Technologies*; Danezis, G., Martin, D., Eds.; Number 3856 in Lecture Notes in Computer Science; Springer: Berlin, Heidelberg, Germany, 2006; pp. 197–209.

116. Wiedersheim, B.; Ma, Z.; Kargl, F.; Papadimitratos, P. Privacy in inter-vehicular networks: Why simple pseudonym change is not enough. In Proceedings of the 7th International Conference on

Wireless On-Demand Network Systems and Services, Kranjska Gora, Slovenia, 3–5 February 2010; pp. 176–183.

117. Hartenstein, H.; Laberteaux, K.P. A tutorial survey on vehicular ad hoc networks. *IEEE Commun. Mag.* **2008**, *46*, 164–171.

118. Lin, X.; Sun, X.; Ho, P.H.; Shen, X. GSIS: A Secure and Privacy-Preserving Protocol for Vehicular Communications.*IEEE Trans. Veh. Technol.* **2007**, *56*, 3442–3456.

119. Buttyán, L.; Holczer, T.; Vajda, I. On the effectiveness of changing pseudonyms to provide location privacy in VANETs. In *Security and Privacy in Ad-Hoc and Sensor Networks*; Stajano, F., Meadows, C., Capkun, S., Moore, T., Eds.; Number 4572 in Lecture Notes in Computer Science; Springer: Berlin, Heidelberg, Germany, 2007; pp. 129–141.

120. Freudiger, J.; Raya, M.; Félegyházi, M.; Papadimitratos, P. Mix-zones for location privacy in vehicular networks. In Proceedings of the First International Workshop on Wireless Networking for Intelligent Transportation Systems (Win-ITS), Vancouver, BC, Canada, 14–17 August 2007.

121. Huang, L.; Matsuura, K.; Yamane, H.; Sezaki, K. Enhancing wireless location privacy using silent period. In Proceedings of the 2005 IEEE Wireless Communications and Networking Conference, New Orleans, LA, USA, 13–17 March 2005; Volume 2, pp. 1187–1192.

122. Caballero-Gil, C.; Molina-Gil, J.; Hernández-Serrano, J.; León, O.; Soriano-Ibañez, M. Providing k-anonymity and revocation in ubiquitous VANETs. *Ad Hoc Netw.* **2016**, *36*, 482–494.

123. Gross, R.; Acquisti, A. Information Revelation and Privacy in Online Social Networks. In Proceedings of the 2005 ACM Workshop on Privacy in the Electronic Society, Alexandria, VA, USA, 7 November 2005; pp. 71–80.

124. Bonneau, J.; Preibusch, S. The Privacy Jungle: On the Market for Data Protection in Social Networks. In *Economics of Information Security and Privacy*; Moore, T., Pym, D., Ioannidis, C., Eds.; Springer: New York, NY, USA, 2010; pp. 121–167.

125. Zheleva, E.; Getoor, L. To join or not to join: The illusion of privacy in social networks with mixed public and private user profiles. In

Proceedings of the 18th International Conference on World Wide Web, Madrid, Spain, 20–24 April 2009; pp. 531–540.

126. Mislove, A.; Viswanath, B.; Gummadi, K.P.; Druschel, P. You are who you know: Inferring user profiles in online social networks. In Proceedings of the third ACM International Conference on Web Search and Data Mining, New York, NY, USA, 4–6 February 2010; pp. 251–260.

127. Thomas, K.; Grier, C.; Nicol, D.M. unFriendly: Multi-party Privacy Risks in Social Networks. In *Privacy Enhancing Technologies*; Atallah, M.J., Hopper, N.J., Eds.; Number 6205 in Lecture Notes in Computer Science; Springer: Berlin, Heidelberg, Germany, 2010; pp. 236–252.

128. Lucas, M.M.; Borisov, N. FlyByNight: Mitigating the Privacy Risks of Social Networking. In Proceedings of the 7th ACM Workshop on Privacy in the Electronic Society, Sydney, Australia, 19–22 May 2008; pp. 1–8.

129. Beato, F.; Kohlweiss, M.; Wouters, K. Scramble! Your social network data. In *Privacy Enhancing Technologies*; Fischer-Hübner, S., Hopper, N., Eds.; Number 6794 in Lecture Notes in Computer Science; Springer: Berlin, Heidelberg, Germany, 2011; pp. 211–225.

130. Guha, S.; Tang, K.; Francis, P. NOYB: Privacy in online social networks. In Proceedings of the First Workshop on Online Social Networks, Reno, NV, USA, 23–24 February 2008; pp. 49–54.

131. Narayanan, A.; Shmatikov, V. De-anonymizing social networks. In Proceedings of the 30th IEEE Symposium on Security and Privacy, Berkeley, CA, USA, 17–20 May 2009; pp. 173–187.

132. Korolova, A.; Motwani, R.; Nabar, S.U.; Xu, Y. Link privacy in social networks. In Proceedings of the 17th ACM Conference on Information and Knowledge Management, Napa Valley, CA, USA, 26–30 October 2008; pp. 289–298.

133. Effendy, S.; Yap, R.H.; Halim, F. Revisiting link privacy in social networks. In Proceedings of the Second ACM Conference on Data and Application Security and Privacy, San Antonio, TX, USA, 7–9 February 2012; pp. 61–70.

134. Takabi, H.; Joshi, J.; Ahn, G.J. Security and privacy challenges in cloud computing environments. *IEEE Secur. Priv.***2010**, *8*, 24–31.

135. Wagner, I.; Eckhoff, D. Privacy assessment in vehicular networks using simulation. In Proceedings of the 2014 Winter Simulation Conference (WSC), Savanah, GA, USA, 7–10 December 2014; pp. 3155–3166.

136. Wagner, I.; Eckhoff, D. Technical Privacy Metrics: A Systematic Survey. *arXiv:1512.00327*, 2015. Available online: http://arxiv.org/abs/1512.00327 (accessed on 5 February 2016).

Chapter 10

Analysis of Intelligent Transportation Systems Using Model-Driven Simulations

Alberto Fernández-Isabel and Rubén Fuentes-Fernández*

Departamento de Ingeniería del Software e Inteligencia Artificial, Facultad de Informática, Universidad Complutense de Madrid, 28040 Madrid, Spain

ABSTRACT

Intelligent Transportation Systems (ITSs) integrate information, sensor, control, and communication technologies to provide transport related services. Their users range from everyday commuters to policy makers and urban planners. Given the complexity of these systems and their environment, their study in real settings is frequently unfeasible. Simulations help to address this problem, but present their own issues: there can be unintended mistakes in the transition from models to code; their platforms frequently bias modeling; and it is difficult to compare works that use different models and tools. In order to overcome these problems, this paper proposes a framework for a model-driven development of these simulations. It is based on a specific modeling language that supports the integrated specification of the multiple facets of an ITS: people, their vehicles, and the external environment; and a network of sensors and actuators conveniently arranged and distributed that operates over them. The framework works with a model editor to generate specifications compliant with that language, and a code generator to produce code from them using platform specifications. There are also guidelines to help researchers in the application of this infrastructure. A case study on advanced management of traffic lights with cameras illustrates its use.

KEYWORDS

intelligent transportation system; smart city; sensor; actuator; traffic lights; simulation; model-driven engineering; modeling language; agent-based modeling; code generation

1. INTRODUCTION

The concept of *smart city* appears as an answer to the challenges posed by the management of highly interconnected physical and Information Technology (IT) infrastructures in relationship with their communities and environment [1]. The goal is to take advantage of the collective intelligence in such complex systems to get more efficient, sustainable, and liveable environments [2]. Intelligent (or Smart) Transportation Systems (ITSs) [3] play a key role in this context. Their services try to improve traffic regarding security, sustainability, and reduced costs in terms of time, money, energy, and environmental impact. Examples of these services [4,5] are the timely update of traffic data, the request of the equipment needed to assist in emergencies using the data provided by vehicles, or the tracking of freight along its journey. For this purpose, these systems make use of the pervasive presence of networks of sensors and actuators (Sensor Networks, SNs) deployed in people, vehicles, and the environment of *smart places*, and the high availability of data regarding participants in traffic and their activities [6].

The development of these complex systems faces important challenges [3,5,6]. First, their potential effects on the environment and living beings make it difficult to set up suitable controlled environments for experiments. Second, these systems usually comprehend multiple distributed components (both hardware and software), frequently embedded in other elements such as vehicles, phones, traffic lights, or roads. This kind of system has high costs of development, testing, and deployment. Third, there are many interacting elements in a real traffic setting, what makes difficult to validate hypotheses on the behavior of ITSs.

Simulation appears as a useful tool to deal with the previous issues [7]. It allows controlling the relevant variables of the problem, running multiple experiments, and the incremental development of ITSs (*i.e.*, gradually replacing models and software implementations by the actual components and environment). However, this approach also presents some drawbacks [8], mainly concerned with the difficulties to guarantee the alignment between the resulting simulation and the initial abstract model.

Model-Driven Engineering (MDE) [9] has been proposed as a way to overcome these limitations of simulations [10]. In a model-driven development, researchers mainly specify their simulations using models compliant with well-defined Modeling Languages (MLs). These models are combined and refined (adding new information or modifying it) until it is possible to generate code from them. Transformations automate some recurrent modifications of models, code, and other related artifacts. In this way, all the information required to produce the simulation is explicitly represented. This facilitates validating artifacts with theories and crosschecking among experiments. For instance, the same theoretical model of drivers' behavior can be the basis for simulations in different platforms, different models of traffic flows in a city can be compared with implementations in the same platform, and all of them can be reused and validated in different projects as long as they keep the same theoretical basis or simulation infrastructure.

This paper presents a MDE framework to develop simulations of ITSs. This framework comprehends: a ML for ITS experts; a *model editor* and a *code generator* tools; and *guidelines* to apply these elements. Figure 1 provides an overview of the framework components and the main related works that influence their development. All considered frameworks adopt MDE approaches, and all their MLs follow the agent paradigm [8,11].

Our ITS ML (ITSML) builds on the foundation of three MLs that address relevant aspects of the problem: the simulation of traffic in [12] (the Traffic ML, TML), and the development of SNs in [13] (the SN ML, SNML), which in turn is partly based on the INGENIAS methodology [14] for agent-oriented software development and its INGENIAS ML (the INGENIAS-ML). The ITSML provides a seamless integration of their concepts (including direct reuse of parts of their MLs) and additional elements. The integration is based on the agent paradigm underlying these languages. Thus, the ITSML distinguishes between *agents* (*i.e.*, intentional and social entities) and *components* (*i.e.*, elements of systems that are not agents).

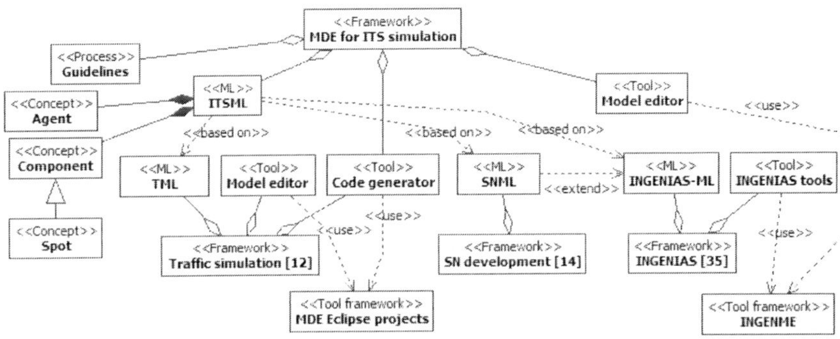

Figure 1. Main components of the Model-Driven Engineering (MDE) framework for Intelligent Transportation System (ITS) simulation (in white) and related works (in yellow). Aggregation relationships (with diamonds) represent whole-part relations, and dependency relationships (discontinuous lines) different relations of use.

Components have a state, are able to produce events, and make available interfaces with methods to manipulate them. Most concepts from the other MLs become *components* here, e.g., the devices of SNs and the elements of the environment in traffic. The execution of methods can be linked in workflows through the production and use of information (e.g., external notifications or internal facts). This ML also introduces the concept of *spot*, as a *component* that indicates where devices can be placed and what they can perceive and act on. For instance, persons, vehicles, and traffic lights are *spots*.

Agents represent both people (e.g., *person* in [12]) and software agents (e.g., *managers* of devices in [13]). They follow a perceive–reflect–act cycle, with specific *components* for each of these tasks. This facilitates decoupling the interaction with the environment and the modification of their internal state.

As general features, the ITSML provides inheritance and instantiation mechanisms. The first one supports the specialization of concept types at the model level. The second one allows defining instances of any type in models, which makes possible the full specification of the simulation at that level. There is also a general association relationship between concepts that can be stereotyped.

The tools of the framework are a model editor developed using the meta-editor INGENME [15], and the code generator from [12]. The first one supports the specification of models compliant with the proposed ML. The second one allows mapping elements in models to code templates that are used to generate the source code of simulations.

The case study that illustrates the use of this infrastructure is the simulation of the management of traffic lights in [16]. That work uses the Simulation of Urban MObility (SUMO) open traffic simulation suite. The paper considers traffic lights that optimize flows in junctions by perceiving the length of their queues and coordinating among them. This case study addresses the development of such simulation using the proposed approach. It shows how most of the simulation can be specified with models and then to generate code from them, which reduces coding effort and facilitates its understanding and later modification. Additional experimentation is also discussed more briefly.

The rest of the paper is organized as follows. Section 2 reviews the background of this work in MDE (see Section 2.1), and the development frameworks for traffic simulations (see Section 2.2) and SNs (see Section 2.3). The framework for the model-driven development of ITS simulations (see Section 3) is described through its ML (see Section 3.1) and its development guidelines and tools (see Section 3.2). Section 4 applies it to the case study of the simulation of intelligent traffic lights (see Section 4.1) and presents additional experimentation (see Section 4.2). The results are compared with related work in Section 5. Finally, Section 6 discusses some conclusions and future work.

2. BACKGROUND

The approach for the development of simulations of ITSs is based on four main works, which are discussed in the following subsections. MDE (see Section 2.1) is the general paradigm for software development used in this research. It is focused on models and their semi-automated transformations. Three development frameworks adopting this paradigm contribute to our research: those for traffic simulations [12] (see Section 2.2) and SNs [13], this extending the INGENIAS methodology [14] (see Section 2.3). The proposed framework integrates elements from these works with specific extensions for ITSs, both at the levels of ML and tools.

2.1. MDE

MDE [9] organizes software development around *models*. These gather most of the information required to produce systems and their related artifacts. The development process is mainly conceived as an iterative refinement of models, where information is added to them and modified. *Transformations* automate some of these modifications, and also generate from models artifacts such as documentation, code, or tests. Examples of transformations can be mapping abstract theoretical models to design ones for a given simulation platform, or adding a default goal of path-following to a driver agent. There are alternatives to work with these elements.

In order to enable the automated processing of models by tools, MDE needs to define formally their MLs. Metamodels are the most widely accepted means for this task regarding graphical graph-oriented MLs, which are the most used [9]. A metamodel defines the primitives available in the language, as well as the constraints applicable to them.

Metamodels are specified using meta-modeling languages. The Meta-Object Facility (MOF) [17] is the standard defined by the Object-Management Group (OMG). It is used to define popular standard MLs, such as the Unified-Modeling Language (UML) [18]. However, tool support for MOF is limited, so researchers and engineers frequently resort to better supported alternatives. Ecore [19], from the Eclipse modeling projects, has extensive tool support and is aligned with the subset of MOF known as Essential MOF (EMOF) [20]. MOF-based MLs use *references* to relate concepts. These references are directed binary links with only a few properties (e.g., super-type and containment). This makes them cumbersome sometimes to specify complex relationships with attributes. The Graph-Object-Property-Role-Relationship (GOPRR) [21] meta-modeling language considers a richer set of primitives for this task. Its languages are organized around *graphs* that contain *objects*connected by *relationships*. Relationship ends are *roles*, and they can appear in any number. All the previous elements can have *properties*. Some tools based on GOPRR are Metaedit+ [21] and INGENME [15].

MDE approaches process information automatically using transformations. Most times, their inputs and outputs are models compliant with metamodels, or text-based artifacts (e.g., code or documentation) compliant with grammars [22]. There are multiple ways of implementing these transformations [23], from modules written in mainstream imperative programming languages, to mappings written in specific declarative transformation languages and executed by engines. Examples of the first approach are wizards in [12] and modules in INGENME [15]; JET [24] and ATL [25] are popular Eclipse-supported examples of the second approach.

The organization of the overall development process also has multiple alternatives. One of the best known is the Model-Driven Architecture (MDA) [26] of OMG. It defines three levels of abstraction for its MLs. Computation Independent Models (CIMs) are intended for domain-specific information, here traffic and very abstract views of ITSs. Platform Independent Models (PIMs) describe general computational abstractions, not depending on specific architectures, e.g., sensors, actuators, and controllers of ITSs. Platform Specific Models (PSMs) include technology-related information, e.g., the target simulation platform. INGENIAS proposes two MDE processes, one based on the Unified Process [27] and another on Scrum [28]. They are oriented to different types of project and team. Both use almost the same kinds of diagram over the entire development, independently of the level of abstraction of the stage. In all the considered processes, designers manually modify development artifacts or run automated transformations on them.

Transformations can also be used to translate information among models of MLs compliant with different meta-modeling languages [29]. This is a less frequent type of transformation, and its specification is quite complex.

2.2. The Traffic Simulation Framework

The *traffic simulation framework* [12] proposes a MDE approach to develop this kind of simulation. It comprehends the TML to describe theories of traffic, and tools to work with it.

The TML is based on Agent-Based Modeling (ABM) [8]. Its agents are intentional entities that represent *persons*. They are modeled in terms of the *goals* they pursue and the *tasks* they can execute to achieve the former. Internally, these agents follow a perceive–reflect–act cycle [11]. An agent perceives its external environment, uses this information to update its internal state, and then to select a suitable goal and a related task according to it, executes this task, and starts again the cycle. This cycle is modeled using *evaluators* (which perform all these activities but the execution of tasks), and *actuators* (that execute the selected tasks).

Regarding traffic, the TML adopts the Driver–Vehicle–Environment (DVE) [30] model. This considers the dynamic interactions among *drivers* moving in an *environment* using their *vehicles*. In these interactions, the mutual influences among the different components affect their features and behaviors. For instance, the mental workload of a driver largely depends on the weather conditions, which are part of the environment. The TML modifies this basic model to be able to integrate different theories of traffic. Its participants are *persons* (that are also the agents), who can drive *vehicles* or not. *Persons* are characterized in terms of the *knowledge* they have and the features of their *profile*.

The ML uses *methods* to specify the actual effects of the interactions among the previous elements. Methods act as placeholders where indications can be provided on how to update dynamically features. There are default methods for several types of entity. An alternative to model these effects is to use the inheritance mechanisms to provide new types of relationship (e.g., *affect* or *impede*), but this can be difficult to specify and semantics not being clear.

Most of the previous concepts can be decomposed in parts (*i.e., components*). For instance, *Profile* has *PComponent*s, and*Environment EComponents*. They can also be extended through inheritance.

The implementation of this framework is based on Eclipse projects. An Ecore metamodel defines the TML, and related tools are based on the Eclipse Modeling Framework (EMF) [19] and the Graphical Editing Framework (GEF) [31]. The two main tools are a model editor and a code generator.

The model editor allows generating graphically instances of the metamodel, *i.e.*, models. These are drawn as graphs of the concepts and relationships included in the metamodel. This editor is directly generated from the metamodel using the EMF and GEF functionality.

The code generator integrates information from different sources to output the source code of simulations. Its inputs include, among others: the traffic metamodel; models compliant with it; the classes EMF generates from the metamodel; libraries describing the target simulation platforms; code templates for specific parts of the generated code; and mappings from model entities to those templates. A template includes text and placeholders that indicate where and what information from models needs to be inserted. The generator manages all this information with graphical wizards. It is implemented using Java with additional libraries.

2.3. The SN Development Framework

The *SN development framework* [13] introduces a MDE approach to develop SNs. It is based on the INGENIAS methodology [14] and its tools for model editing [15], and Eclipse transformations [24,25].

INGENIAS is a MDE methodology to develop Multi-Agent Systems (MASs). MASs are a general paradigm (not only for simulations) where the basic abstraction is the *agent*. As it happens in ABM, INGENIAS agents are intentional and social entities. They execute *tasks* pursuing certain *goals*. These *tasks* can change the internal state of the *agent* or its external environment, or trigger interactions. The environment is modeled using *environment applications*, which produce *events* and have *methods* to act on them. Sometimes, these *events* trigger *tasks*. Interactions exchange information among *agents*.

The INGENIAS-ML supports limited forms of inheritance and instantiation relationships for agents. The latter defines the actual *agent* instances that exist in the system from their types. This is used, for example, to specify the system deployment.

The INGENIAS tools are currently based on the INGENME framework [15] for MDE. INGENME supports the definition of MLs based on its implementation of GOPRR [21]. Its meta-editor allows generating functioning model editors from metamodels. These editors are similar to those of GEF [31], but simpler to describe, as they involve less intermediate models. Additional modules can be integrated in the basic version of these editors. These modules are implemented in Java, and have access to programming interfaces provided by the framework to manipulate both metamodels and their models. Among other tasks, they implement transformations. Text-generation modules are based on templates, as in the traffic simulation framework (see Section 2.2).

The SN development framework extends the INGENIAS-ML to model SNs with the SNML. Its models are organized around *containers*. *Containers* are computational devices where other *resources* run. *Resources* can be pure software *utilities* or*devices* with hardware components (*i.e., sensors* and *actuators*). Communication *channels* linking *containers* are considered a particular type of *resource*. The control of these *resources* is the responsibility of *managers,* which is a role that *actor* agents play. These *managers* can be *passive* or *active controllers*. The first group only listens to notifications from *resources,* while the second group can also invoke their modification methods.

The framework includes basic guidelines on how to model with the SNML and translate these specifications to the INGENIAS-ML. The rest of its development follows an INGENIAS process [27].

3. THE MDE FRAMEWORK FOR ITS SIMULATION

The framework for the development of ITS simulations comprehends the following elements: a ML (see Section 3.1), and guidelines to apply it with a model editor and a code generator (see Section 3.2).

3.1. Modeling Language

The ITSML is organized around the main concepts that appear in this kind of system and its context (see Figure 2). There are *persons* who interact with *vehicles* and an external *environment*, which is composed of *things*. All these elements are *places*that can contain multiple *spots*. A *spot* is an element that can be observed and/or where *containers* can be located. *Containers*are computational nodes, linked by communication *channels*. In them, software *managers* control the deployed *sensors* and*actuators*. All the previous concepts share some features (e.g., they are modeled with a state and an access interface with methods, and generate events). A common abstract super-concept named *component* gathers these features.

That conceptual basis is the result of integrating and extending multiple researches related to the domain. The ML follows an ABM approach [8], so it makes of agents, and their interactions among them and with external objects, the core of modeling. This perspective is common to several MLs that provide concepts for the ITSML. The TML [12] is the main one, given the close relationship between the domains of traffic and ITS simulations. The INGENIAS-ML [14] offers a rich set of primitives for ABM and basic extension mechanisms of limited inheritance and composition. It is also the foundation of the SNML [13], which models concepts of devices and their controllers in SNs.

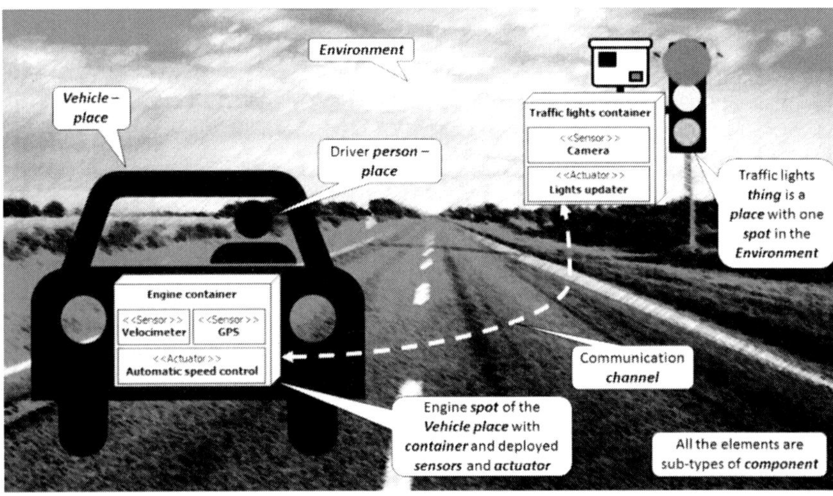

Figure 2. Main concepts (in *italics*) of the ITS Modeling Language (ITSML) in context.

The ITSML specifies that conceptual framework using a metamodel. Working with metamodels allows reusing parts of the other MLs. For instance, it supports introducing and modifying concepts and relationships as specializations of those present in the different base metamodels. There are also extensions that mechanisms in the base MLs do not support, but those of metamodels do. For instance, the ITSML has to allow defining instances of arbitrary types in its models.

The meta-modeling language used for the ITSML is the implementation of GOPRR [21] in the INGENME framework [15]. In next diagrams of this section, nodes and relationships, respectively, represent meta-entities and meta-relationships. The ends of these meta-relationships are the meta-roles. Meta-properties appear as attributes or adornments of the previous elements. Examples of these are the numbers in the meta-roles that are cardinality indications. Meta-relationships with triangles represent inheritance, and those with diamonds aggregation, with filled diamonds representing composition and hollow diamonds mere aggregation. This notation is similar to that of UML [18]. As the ITS metamodel integrates concepts from different metamodels, these have been located in packages. Names of concepts and relationships from other metamodels appear qualified with the name of its package, *i.e.*, *traffic* [12], *SN* [13], or *INGENIAS* [14].

The rest of the section discusses in detail the concepts for ITSs in the metamodel. It starts with the general *component* (seeSection 3.1.1). Then, it focuses on the *places* where *sensors* and *actuators* can be deployed and with which they interact (seeSection 3.1.2). *Managers* control these resources to provide services to *persons* (see Section 3.1.3) moving in an *environment*(see Section 3.1.4). There are also some general mechanisms applicable to all these concepts (see Section 3.1.5).

3.1.1. The *Component* Entity

Figure 3 is focused on the *component*, which is the base concept in most of the inheritance hierarchies of the ITS metamodel. The SN *device* and the constituents of the traffic *person* (with the exception of the mental entities *goal* and*knowledge*), *vehicle*, and *environment*, become *components*. *Components* have an internal state and an interface composed of *methods*. *Methods* can be *external* (when they can be invoked from outside their *components*), or *internal* otherwise.

The state and methods are characterized in terms of information. Pieces of *information* can be of three types: *facts* make up the internal state; *events* are information that appear in *components* but they do not explicitly produce, and that can be used internally or externally; *notifications* are information produced for external use. *Methods* output both *facts* and *notifications*, but *events* come from non-specified parts of *components*. Besides their results, methods are defined by the parameters they need and the conditions that enable them, which can involve any kind of *information*. *Components* can be connected in pipelines using the *information* their *methods* produce and consume.

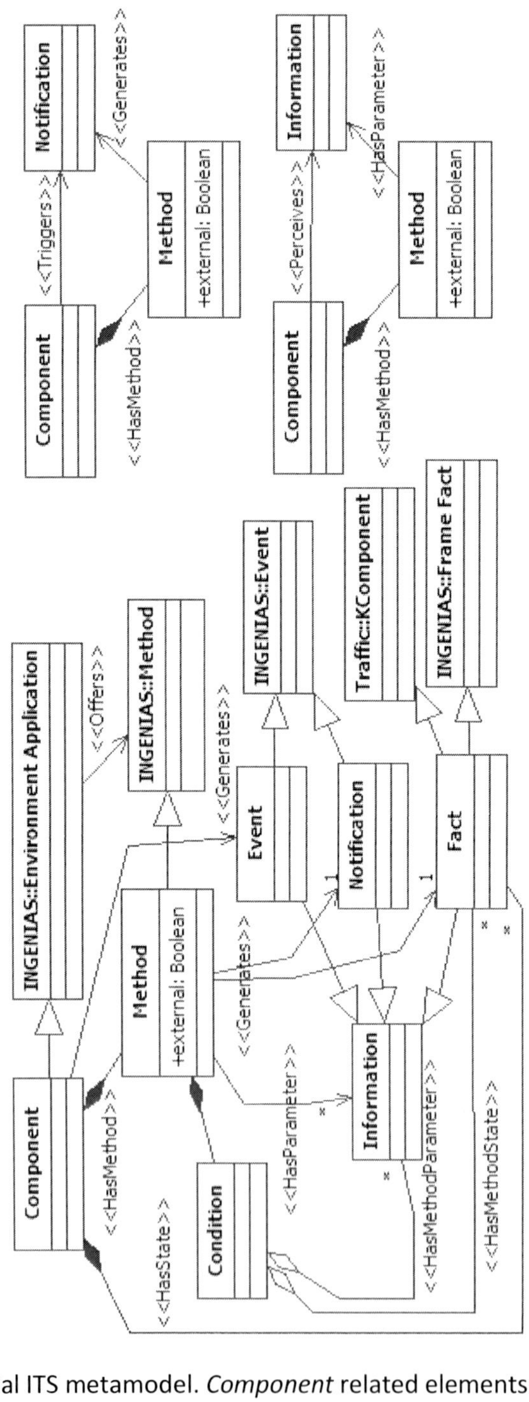

Figure 3. Partial ITS metamodel. *Component* related elements.

The metamodel defines two convenience meta-relationships for *methods* and their related information: *triggers* and*perceives*. The first one relates a *component* to a *notification* that one of its *methods* generates. The second one links a *component*to any *information* that is input to one of its *methods*. These meta-relationships allow focusing on the considered information and omitting the particular *methods* involved.

3.1.2. *Place* and *Device* Related Entities

An ITS needs to gather information from its surroundings and act on them. Figure 4 shows the high-level structure of this context. It is abstracted as a set of *places* that contain multiple *spots*. These *spots* are the *components* that can be actually monitored and modified, and where devices can be placed. Following the DVE model adopted by the TML (see Section 2.2), the participants in the traffic phenomena (*i.e.*, *persons*, their *vehicles*, and the *things* in the external *environment*) are *places*. Their components according to the TML become *spots*.

Following the SNML (see Section 2.3), *devices* are responsible of sensing and acting in the ITS surroundings. The metamodel incorporates the device-related concepts from SNs as *components*. A *container* is the computational resource physically attached to a *spot*. Communication *channels* link them. A *container* runs multiple *devices* (*i.e.*, *sensors*, *actuators*, and*utilities*). Both *sensors* and *actuators* can invoke consult *methods* of *spots*. *Sensors* can also perceive their *notifications*, and*actuators* invoke their external modification *methods*. *Utilities* are pure software components without access to the surroundings.

Sensors and *actuators* are thus related to two *spots*, which can be (or not) different: the one to which their *container* is attached (*i.e.*, *HasAttached* meta-relationship); and the one with which they interact. For instance, a *place vehicle* has a *spot* for a *container* with a *thermometer sensor* that measures the *air* temperature, being that *air* a *spot* in the *place weather conditions* of the *environment*. The *perceives* and *actuates* meta-relationships are convenience meta-relationships for these interactions. They link *sensors* and *actuators* with the *spots* they manipulate without considering the involved *methods*.

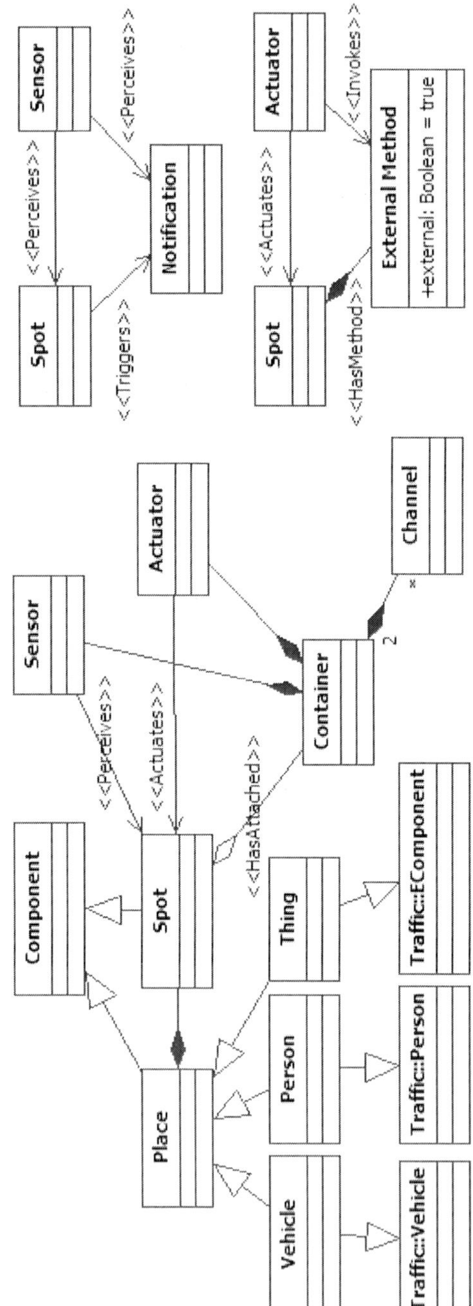

Figure 4. Partial ITS metamodel. *Place* and *Device* related elements.

The control of *devices* can be modeled in their *methods* or using *manager agents*. The activation of a *method* happens when its inputs (e.g., external *notifications* and internal *events* or *facts*) and the internal state of its *device* meet some *conditions*. As a consequence, it generates new information, either internal to the *device* (as *facts*) or external for other components (as*notifications*). For instance, a *timer sensor* in a traffic light has a *method* activated by a *time-consumed event*. When this happens, this *method* produces a *change notification* for the *actuator* in charge of changing lights, which processes it with another*method*. This *method* changes to a target light that depends on a *fact* that represents the current light, and which is part of the internal state of that actuator. On the other hand, *manager agents* (see more on them in Section 3.1.3) control *devices* when complex algorithms are needed (as they have a rich model for decision-making) or there is communication among *containers*(which is one of the specific responsibilities of agents).

3.1.3. The *Person* and *Manager* Entities

The ITS metamodel models with *agents* complex intentional and social components. These include *persons* and *managers*that control *devices* (see Figure 5). The specification of *agents* includes the information they have, their goals and capabilities, and the process to manage the previous elements.

Knowledge specifies the internal state. It includes *facts* (*i.e.*, any statement about the world) and *goals* (*i.e.*, states of the world that an *agent* wants to keep or to achieve).

Tasks are the high-level actions of agents. They are implemented using *methods* or other *tasks*. When using *methods*, a *task* is eligible for execution depending on the *condition* of the *method*. In this case, the *task* can change the *agent*'s *knowledge*, use*devices*, or act on the environment through *notifications*. Communication among agents is modeled as a particular use of*notifications*. The alternative to define the execution of a *task* is as a workflow of interconnected *tasks* to set up complex activities. These workflows follow the INGENIAS-ML [14].

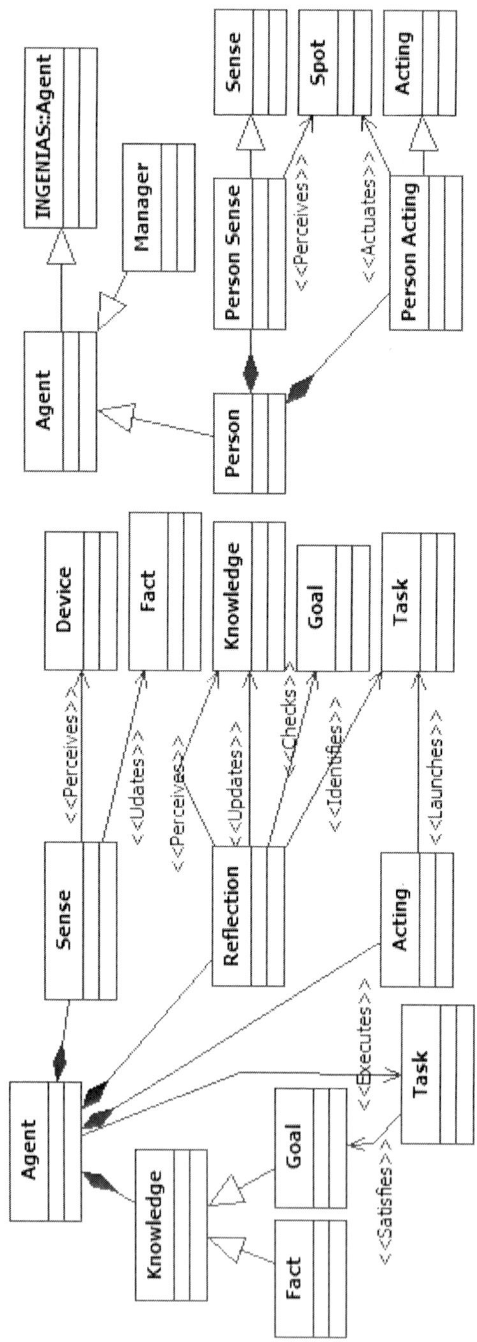

Figure 5. Partial ITS metamodel. *Person* and *Manager* related elements.

Agents work in a perceive–reflect–act cycle that defines its dynamic behavior. This cycle is modeled with three types of*components* (*i.e., sense, reflection,* and *acting*) that operate over their own and their *agent* internal state. The *sense* meta-class models perception as an interaction with observed *devices* that adds or modifies *facts* in the state of the *agent* or the *sense*. The instances of the *reflection* meta-class are responsible of the reflection stage. They analyze available *knowledge* (both *facts* and*goals*), and produce new one or modify the existing. Then, they check non-satisfied *goals*, and identify the *tasks* related to them and suitable for execution in the current context. Finally, the *acting* meta-class represents the actual execution of one of the executable *tasks*.

The *agent* concept is suitable to represent directly control *managers*. For *persons* involved in traffic, there are some modifications. A *person* has specific *person sense* and *person acting components* that allow interacting directly with *spots*without the need of intermediate *devices*.

The previous definitions allow aligning concepts from the ITSML with those of its base languages. The ITSML *agent*, *fact*, and *goal* are sub-types of the INGENIAS-ML meta-classes *agent*, *frame fact*, and *goal*. Both the TML *knowledge* and *profile* are modeled as ITSML *knowledge*, and the TML *goal* is a super-type of the ITSML *goal*. The *person* sub-type of *agent* is also a subtype of the TML *person*.

3.1.4. The *Environment* Entity

Regarding the *environment*, it represents the elements participating in traffic that are not people, their vehicles, or the ITS components (e.g., *devices* and their *managers*). Examples of its elements are the roads, the traffic signals, the weather conditions, or obstacles. It is an extension of the TML *environment* (see Figure 6). It is modeled as an aggregation of *things*, which are located over a *map*.

The *map* represents the distribution of the physical space. It is a kind of graph composed by a set of *section*, each one linking two *junctions* (at least one) and having a given *length*. This length can be measured in different units, e.g., kilometers or number of vehicle along it. A *section* is a set of *lanes* that can be direct (from the origin to the destination junction) or reverse (the other way). A *lane* has an *occupants* list of references to the vehicles or persons in it ordered according to their position.

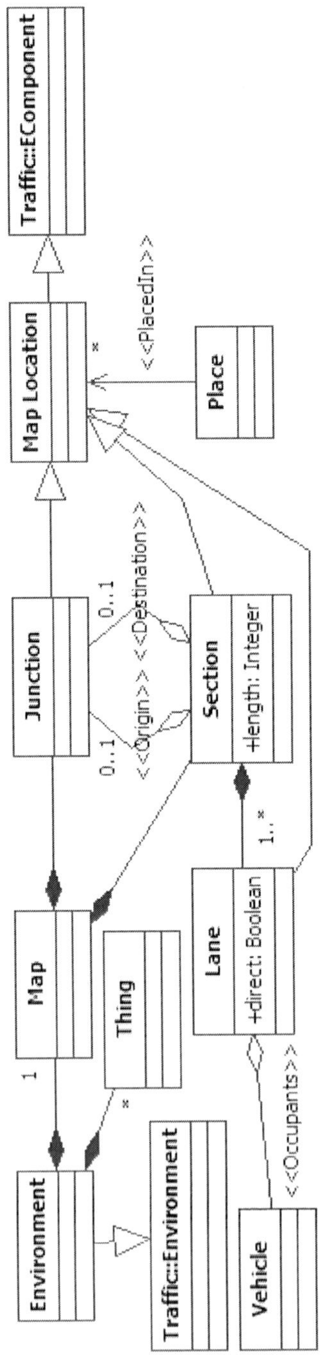

Figure 6. Partial ITS metamodel. *Environment* related elements.

The other elements of the environment (the equivalent of the TML *EComponent*) are *things*. They can be placed in specific*map locations*, but also being considered global for the environment, e.g., the weather conditions or the sun position.

Vehicles and *persons*, being *places*, can have a location. In the case of instances of *vehicle*, this is mandatory. *Persons* can have a location, when they are pedestrians, or being attached to a *vehicle*.

3.1.5. General Features

The ITS metamodel also includes general instantiation, inheritance, and association mechanisms. For each meta-class of name *name*, there is another meta-class of name *Iname* that represents its instances at the model level. An instance of the later can be linked to an instance of the former with an instance of the meta-relationship *instanceOf*. For example, there is a meta-class *IPerson* linked by a meta-relationship *instanceOf* to the meta-class *Person*, so in a model, a specific driver *IPerson* can be defined as an *instance of* a specific type of *Person*. The inheritance mechanism extends those available in the traffic, SN, and INGENIAS MLs. It can be applied to most of concepts in the ITS metamodel, with the semantics that the sub-concept inheriting from the super-concept has all its features, but it can extend or constrain them with additional elements. Default constraints establish that an animated concept (*i.e.*, an instance of the meta-class *person*) cannot inherit from a non-animated concept (*i.e.*, an instance of a meta-class that is not *person*), and *vice versa*. Regarding association, there is a general meta-relationship *relatedTo* between instances of *component*. It has a *stereotype* property to establish arbitrary types of it.

3.2. Development Guidelines and Tools

Figure 7 shows the development guideline for the framework. The process has two different stages. The first one (see nodes 1–20) is specific of this work and focused on modeling with the ITSML. The definition of its metamodel (see Section 3.1) provides hints to identify the different concepts. The results of this stage are models that provide an abstract representation of the ITS and its environment. They are independent of specific target simulation platforms. The second stage takes these models as input and runs transformations to get models with the INGENIAS-ML (see activity 21), and then applies an INGENIAS development processes [27,28] to get the simulation code (see activity 22).

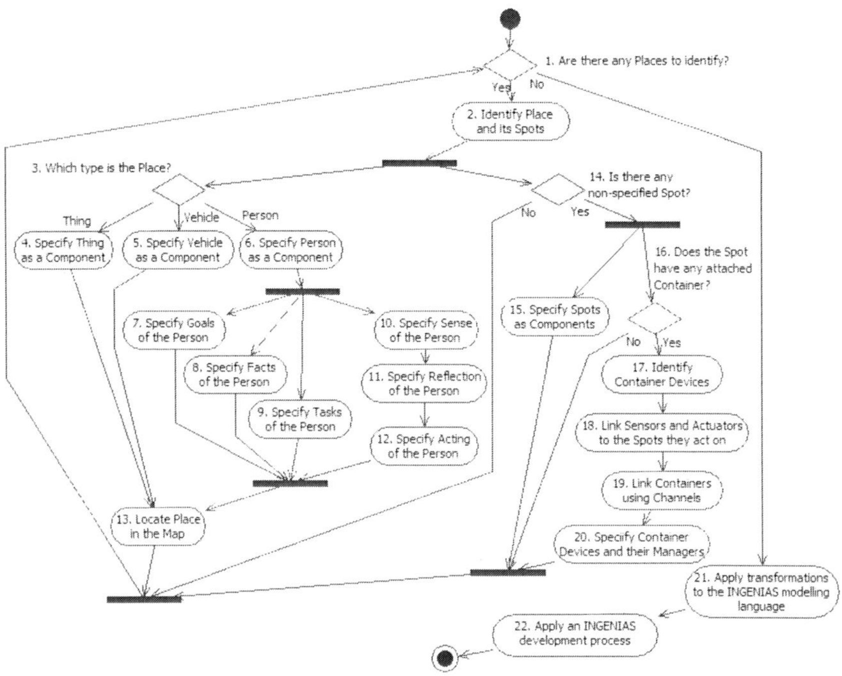

Figure 7. Activity diagram of the development process with the ITSML. Concepts starting with uppercase belong to this ML.

The process starts identifying *places* in decision 1. These establish the context of the system and the elements over which the ITS must work. If there is any *place*, activity 2 assigns it a name and identifies its *spots*. Next activities in this path specify the *place* as a *component* (starting with decision 3) and its *spots* (see decision 14). If there are not more *places* to identify, the abstract modeling of the ITS has finished, and the process starts the INGENIAS development (see activities 21 and 22).

The description of a *place* depends on its type (see decision 3): *thing*, *vehicle*, or *person*. All of them have to be specified as*components* (see activities 4–6). This specification indicates the information available from these elements, and the means to interact directly with them. Instances of *person* should not usually include any external *method* to modify its internal knowledge, as it would violate their autonomy according to the agent paradigm. Other of their features can be modified from the environment, for instance to reflect that they have suffered some injury. If needed, these *places* must also be located in the *environment map* (see activity 13), which also implies specifying this.

The specification of a *person* also comprehends its behavioral cycle. This is specified in terms of its *sense* (see activity 10),*reflection* (see activity 11), and *acting* (see activity 12) *components*. *Sense* instances use their *methods* to get *notifications* from external sources (*i.e.*, *spots* and *devices*), and include it as *facts* that are part of the internal state of the *person*. *Reflection*instances model several activities: generation of new derived *knowledge* from that already available; checking *goal*satisfaction according to the current internal state; selection of potential *tasks* to execute according to the state of *goals* and other *knowledge*. *Acting* instances represent the actual execution of *tasks*. Activity 7 is used to specify those *goals*, activity 8 the *facts*, and activity 9 the *tasks*. These elements are linked among them: *tasks* are potentially able to satisfy some *goals*; *goals*are satisfied according to conditions specified in terms of *knowledge*; and *tasks* are eligible depending on the state of associated *goals*.

After specifying a *place*, the process continues with its *spots*. These *spots* are elements where *devices* can be located, or that*devices* can observe and act on. The specification of *spots* starts identifying them (see decision 14). Every *spot* must be defined as a *component* to determine its related *information* and *methods* (see activity 15). *Spots* are also the elements that can have attached *containers* (see decision 16). Activity 17 identifies the *containers* attached to a *spot* and the *devices* running in it.*Containers* are connected through the *channels* identified in activity 19. Among *devices*, *sensors* observe *spots* and *actuators* act on them. These *spots* can be or not the same to which their *containers* are attached. Activity 18 links *sensors* and *actuators* to their target *spot*. Finally, activity 20 specifies the *container devices* as *components* and their *managers* as *agents*, if they exist. Although not included in the figure, the specification of *managers* should follow a similar workflow to that of *person* (see activities 6–12), as both entities are sub-types of *agent*.

Activities 21 and 22 correspond to the transition and the development with an INGENIAS process [27,28]. The first activity develops, if not available, and runs an automated transformation with input ITSML specifications and output INGENIAS-ML models. Then it follows an INGENIAS development process to generate the simulation source code. The main issue here has to deal with code generation. It relies on platform-specific templates (see Section 2.3), which must be available from other projects or need to be developed from scratch.

This approach is in line with MDA [26]. The ITSML models are CIMs and PIMs, and the INGENIAS-ML models are PIMs and PSMs, in both cases depending on their level of abstraction.

Figure 7 shows only a sequential perspective of the process. In fact, its actual application needs to be iterative in its different paths. For instance, in the path from nodes 3 to 13, activities 4–6 can be done in the first iterations to give a high-level description of the system context, while the other activities correspond to a later and more detailed specification. In that path also, when modeling the *person tasks* (see activity 9), it is possible to find out that their execution requires additional internal *methods* of *person*, so activity 6 must be revisited. Moreover, when in activity 18 designers link *sensors* and*actuators* to the *spots* they work on, this can lead to modifications in the definition of those *spots* (made at some moment in activity 15), or even of their *places* (starting in decision 3).

The previous process is supported by two main tools. A graphical model editor allows generating model instances compliant with the ITS metamodel. This editor has been developed using the infrastructure of INGENME [15]. A code generator takes as input the metamodel and the model instances to generate the source code of the simulation for a target platform. Researchers only need to provide the code templates for that platform, and give the correspondences between those templates and the concepts of the ITS metamodel. The code generator has been developed for the traffic simulation framework [12]. It replaces here the modules used in INGENIAS, which offer less assistance in this process.

4. EXPERIMENTATION

This section illustrates the application of the previous framework (see Section 3). It includes a case study (see Section 4.1) and a report on additional experiments (see Section 4.2).

4.1. Case Study: A Control System for Traffic Lights

The availability of affordable sensors and actuators for ITSs has made traffic experts reconsider some of the traditional designs of traffic facilities. In particular, traffic signals can be made more *intelligent*, in the sense of being aware of the state of traffic in their areas and coordinate with other signals to improve the overall state in wider areas. The use of *intelligent*traffic lights to control junctions in streets is part of this approach [16]. In that work, the *intelligence* comes from two main features of these signals. First, they are able to perceive the number of cars in the lanes that join in their junctions. They achieve this using Optical Information System (OIS) sensors. Second, they exchange information to adapt their behavior to the overall traffic state in the area with the goal of maximizing the throughput of cars from it. In order to evaluate the benefits of the use of these intelligent traffic lights over traditional ones, the work in [16] simulates both situations using the SUMO platform and real data of traffic. This case study reproduces this experiment using the ITS simulation framework.

The process starts with the identification of potential *places* in the problem (see decision 1 and activity 2). According to its definition (see Section 3.1), a *place* is the component that contains *spots*, which are the elements where *containers* can be attached or that *sensors* can monitor and *actuators* act upon. The presentation of the problem introduces several of these*places*. *Junctions* contain intelligent traffic lights where OIS sensors are located. These sensors perceive lanes from *streets*.

At this point, the process can continue by specifying those *places* (starting in decision 3) or the *spots* and related *devices* (starting in decision 14) (see Figure 7). As the original work pays particular attention to the system *devices*, this case study continues in the second path. Figure 8 shows the model instance resulting of it.

Decision 14 begins identifying the *spots* of the previous *places*. The traffic lights are considered as located in the *junction posts* surveying the main *street sections*. Both *junction posts* and *street sections* are *spots*, as the former represent where *devices*are located, and the latter what they survey and affect. Activity 15 specifies these *spots* as *components*. In this case, they do not have any specific state or functionality to specify. Only *street sections* could have it related to the vehicles in them. However, this specification is considered later in the environment *map*.

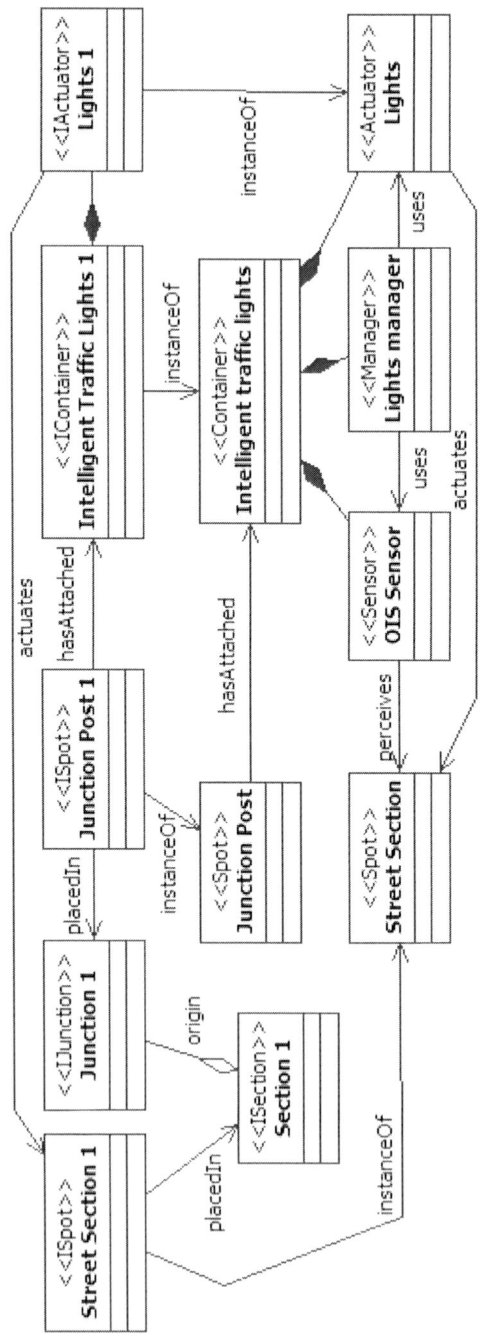

Figure 8. Intelligent traffic lights. *Container* related elements.

The next nodes in the path focus on the *devices* the simulated ITS uses to perceive and act. The first step in decision 16 is discovering the *spots* with *containers* for these *devices*. In this case, there is only one of such *spots*: the *junction posts*.

Activity 17 identifies the *devices* running in the *containers*. The *intelligent traffic lights* are modeled as *containers* in *junction posts*. They have as *sensors* the *OIS sensors* that allow them perceiving the queues in their lanes (using the *street sections*); their actuators are their red, green, and yellow *lights*. Activity 18 links *OIS sensors* to the *street sections spots* they watch, and *lights*to the *street sections spots* they are pointing at.

The *containers* of ITSs are not isolated. They set up a complete system thanks to its communications through *channels*. Activity 19 identifies these channels. The *intelligent traffic lights* are connected in this way (not shown in the figure) to share the state of the whole area.

Activity 20 models the *sensors* and *actuators* of *containers* as *components*, and their *managers*. An *OIS sensor* has a method to consult the number of vehicles in its lane. A *light* has a state characterized by its current color, the time remaining in it, and the current default time assigned to each color; its methods allow checking the previous values and changing the periods assigned to each color and which to display. As in the SNML, complex control algorithms are modeled using the *managers* of*devices*, which are agents. In [16], traffic lights adapt the time assigned to each color according to the information from their sensors and from traffic lights in the other junctions. For this problem, a *lights manager* orchestrates work in each *intelligent traffic lights* container. It obtains information from its *OIS sensors* and sends it to the *managers* in the other junctions using the*channels*. It also receives from the *channels* the updates from the other *junction posts*. Finally, with all of this information, it updates the lights state.

Figure 8 also shows how to introduce instances of the types identified for a given problem. The instance *lights 1* of the class *lights* runs in the instance *intelligent traffic lights 1* of the *container*, and it governs the traffic of the instance *street section 1*of the class *street section*.

After modeling *spots*, *containers*, and their *devices*, the specification needs to consider the persons and their vehicles, as well as the environment where all these elements interact. This corresponds in the process to the paths for *places* starting in decision 3 (see Figure 7).

In [16], only vehicles are modeled, as drivers are implicit in their control. Using data from real places, researchers determined the volume of traffic in the different sections of streets and the junction turning percentages. In the ITSML, the*vehicle* is a passive *component*, and the decision-making is modeled with the driver *person*.

Activity 5 models the *vehicle component*. In this case, it has no state. The *environment map* models its position (see later activity 13). Its methods should move the car, change the lane, and check its own position, that of other vehicles around it, and the color of the traffic lights in its *street section*.

The specification of *person* is more complex. It includes the elements representing its basic knowledge and actions (see activities 6–9), and the perceive–reflect–act cycle (see activities 10–12).

Activity 6 specifies the *person* as a *component*, mainly regarding the external methods intended to check its state. Here, no one is needed. Activities 7–9 model its knowledge and capabilities. The *goal* of the driver is the path to follow in the street. It is characterized by the entry and exit points to the street, and the exit to take at each junction. The agent only needs to know the position of the vehicle to make decisions. This is a *fact* that directly translates the information provided by the *vehicle*methods. The *tasks* of the agent are those to check the environment and manoeuver the *vehicle*. In the first group are *checking junction* (to know when to turn and in which direction), *checking position* (to know if there is a *vehicle* in a given position), and*checking lights* (to know if the *vehicle* can go into the junction given the traffic lights). The second group includes *moving forward*, *turn* (left or right and the number of exit), and *stop* the *vehicle*.

Activities 10–12 use the previous elements to represent the acting cycle of the driver *person*. Activity 10 models *sense*. It only needs to *perceive* the *vehicle spot* to assert the *facts* regarding its current position, whether the surrounding positions are busy, and the color of traffic lights. Activity 11 models *reflection*. It compares the current position with the next turn. If the*vehicle* has not reached the turning position and lights allow it, it chooses the *task* to move forward; if it has reached the position, it chooses to turn when the target position is available; in other case, it waits. Activity 12 models *acting*, that translates one of the potential *tasks* for execution to *person* and *vehicle methods*.

The last *places* to describe are *things* in the environment. Activity 4 specifies them as *components*. The already mentioned*junctions* and *streets* are *things*. Here, their specifications are empty because these *places* are only containers for *spots*.

The final step in this modeling is activity 13, where *places* are located in the *environment map*. In [16], there are three junctions across a street, and each junction connects two sections of the street and a transversal section from other streets. The *thing street* is located over the three junctions and two sections. These elements and the transversal sections have direct correspondences in the ITSML as parts of the *map*. The model also needs to specify the number and orientation of *lanes* in each *section*.

Figure 9 shows the model resulting of these activities, omitting *lanes* of *sections* and the links between the *street* and*junctions*. All the elements appear as instances (*i.e.*, stereotypes *Iname*) of classes (*i.e.*, stereotypes *name*) that are the direct representation of their meta-classes (*i.e.*, *name* of the meta-class). For instance, the metamodel (see Section 3.1) includes a*section* meta-class. *Section 1–2* and *section transversal 1–3* are instances of a class *section* that is a direct instance (without any modification) of the meta-class *section*. The reason of this modeling is that specifying these instances only needs using existing elements, and not adding new state, methods, or relationships.

When all the previous modeling has finished, activity 21 runs the transformations to create INGENIAS-ML models from these ITSML specifications. Their correspondences follow the inheritance relationships of meta-classes in the ITSML to those of the INGENIAS-ML when they exist. For instance, the ITSML *components* in models become INGENIAS-ML *environment applications*, as *component* is a sub-meta-class of *environment application* (see Figure 3). In the case of multiple inheritances, the base target meta-class is that with the closer semantics. For instance, the ITSML *person* inherits from the INGENIAS-ML*agent*, and through *place* and *component* also of the INGENIAS-ML *environment application* (see Figure 3 and Figure 4). However, its meaning is closer to the INGENIAS-ML *agent*. Thus, the transformation generates for a *person* an INGENIAS-ML *agent*, though it also adds additional elements corresponding to it being also a *place*.

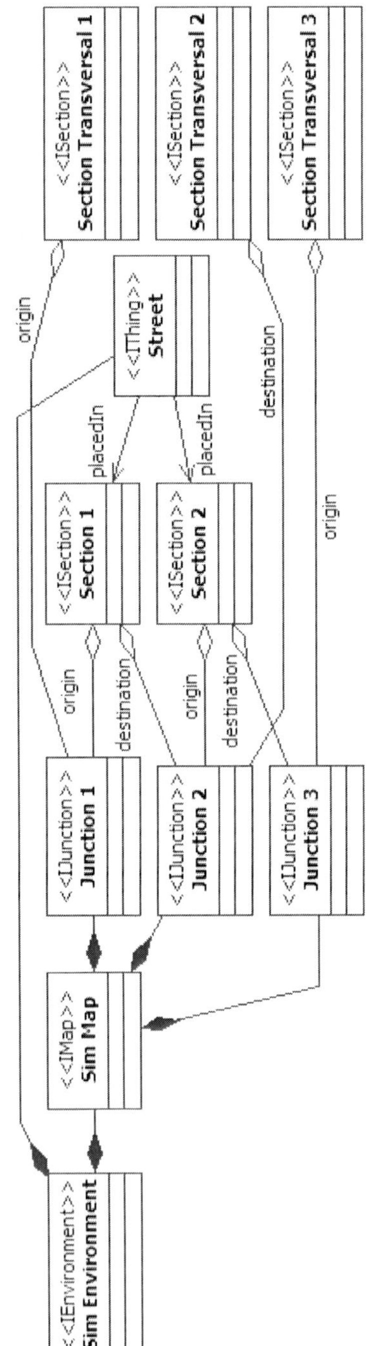

Figure 9. Intelligent traffic lights. *Environment* related elements.

Finally, activity 22 follows an INGENIAS process [27,28] to refine those INGENIAS models and generate the source code for the simulation in SUMO [16]. In this case, INGENIAS models mainly add the description of the interactions among*manager* agents regarding the exchange of information from their *intelligent traffic lights*. SUMO specifies most of the components of its simulations using XML files. The main mappings with the original concepts from the ITSML are the following. The *map* of the *environment* is represented in SUMO as a *network*. Its *nodes* and *edges* are, respectively, ITSML*junctions* and *lanes*. The SUMO *vehicle* merges the ITSML *vehicle* and driver *person*. SUMO does not allow controlling the movements of its *vehicles*, as it provides only a limited set of implementations for them. Nevertheless, the path vehicles have to follow can be specified, and SUMO controls that traffic signals are observed. Regarding traffic lights, SUMO supports*actuated traffic lights* for the experiments described in [16]. They can change the time showing each color according to the number of vehicles in the *edges* they control.

4.2. Discussion

The proposed framework has been tested in several experiments with the purpose of assessing its components. These experiments considered different contexts in order to evaluate its adaptability to multiple needs regarding simulation. In particular, the ITSML has been used to specify different models for people and types of ITS, and the process and tools also for several target platforms.

Regarding decision-making (see Section 3.1.3), there has been work with the goal-model of INGENIAS [14] and the reactive model for drivers in [32]. Compared to the ITSML, INGENIAS offers additional primitives to model goals with the distinction between hard (with quantitative evaluations) and soft (with qualitative evaluations) goals, and different relationships (e.g., affects, contributes positively, satisfies, and AND and OR compositions). The work in [32] offers an alternative perceive–reflect–act cycle. Its agents consider timed data, and combine for reflection a reactive model and a short-time planner.

The concepts in the previous models are specializations of types already available in the ITSML. For instance, hard and soft goals [14] are sub-types of *goal*, and the short-time planner [32] of *reflection*. Relationships can be represented using instances of the *relatedTo* meta-relationship.

Nevertheless, this is a limited approach, as it does not allow introducing constraints. For instance, INGENIAS relations between goals can be modeled, but not constrained with graphical primitives to link only goals.

There is also work with other types of ITS. Smart roads [33] are ITS which SNs are mainly deployed on roads. Their models with the ITSML use specializations of concepts related to devices and managers (see Section 3.1.2) and the environment (see Section 3.1.4). In [33], there are eye cats (a type of *spot*) deployed on road borders (which are *things*). The eye cats hold magnetic trackers (as *sensors*) that perceive vehicles passing near them. *Manager agents* represent the control of these sensors and their communications.

The framework has also been tested with different target simulation platforms. Besides the already mentioned SUMO [16], it has also been tested with MATSim [34] and the platform for traffic simulation in [35]. MATSim mainly defines vehicle movement using paths and additional parameters that influence it, as SUMO does. Thus, the approach described for SUMO in the case study (see Section 4.1) is applicable with minor changes to MATSim. MATSim can also be extended with additional behavioral modules, as the platform in [35]. The latter distinguishes between vehicles and drivers, and supports complex agent-based models for people. Drivers have a basic behavior of path following that can incorporate maneuvers and general constraints on movement as goals. With the ITSML, those constraints are also modeled as *goals*, linked to *tasks*that represent maneuvers. Tasks are related to *conditions* that represent when those maneuvers can be considered for execution. In all the cases, the first tests with these platforms require developing specific code snippets to connect already available code (that for abstract models) with the platform. This also happened in the experiment reported in the case study. These new fragments could later be reused, or adapted with less effort, when experiments continued, and integrated in the code generation tasks.

In these cases, the process points out the ITSML primitives suitable to model the concepts from those models and platforms. However, two aspects remain little described, the choice of relationships and the use of inheritance. Relationships are deduced from the context provided by concepts. The process focuses on the latter, and these determine the allowed relationships. There are no indications on how to choose when several relationships are available beyond their description in the ITSML. There are also sometimes difficulties to design inheritance hierarchies, particularly to establish what concepts must be extended.

5. QUALITATIVE COMPARATIVE ANALYSIS

This work is related to two main aspects of research in ITSs. It deals with the relevant concepts to model this kind of system, and the approaches followed in their analysis, development, and simulation.

ITSs cover a wide variety of needs [3,5,6,36], e.g., reduction of traffic congestions, toll collection, parking, collision avoidance, or emergency response. However, their architectures and perspectives on the environment present a similar structure. There is a world external to the system that includes, for instance, environmental conditions (e.g., weather, sun position, or daytime) [33,37], roads [33,38], vehicles [37,39], or pedestrians [39]. ITSs include devices (sensors and actuators), which are their "interface" to interact with that environment [6,40]. Using them, ITSs provide their services. These services may require computational resources beyond devices, e.g., servers to store information or allow users to monitor the system and the environment.

The previous facets can be effectively integrated using a view of these elements based on external interfaces able to manipulate their state. In most cases, their inner structure and behavior can be disregarded. For instance, with the exception of works focused on device control (e.g., [39,41]), most of research adopts a high-level view of devices, focused on the data they gather and the services they provide (e.g., [33,38,40]). Such perspective is also applied in the models of environment elements when needed. The only exception is the map of the environment, which is modeled with some kind of graph, e.g., when working with location information from the Global Positioning System (GPS) [38].

The point of view of simulation is slightly different. It needs to integrate concepts to model the objects in the external reality and their behavior, so it can represent their influence in sensors and actuators [42]. For instance, drivers' behavior is not only an aspect to observe, but also to manipulate in order to test the person and system functioning in multiple circumstances.

Regarding vehicles, simulations need at least to establish partially their path (*i.e.*, origin and destination, and sometimes intermediate points). The specific way in which a vehicle meets this objective offers variations. There are works that consider an strict adherence to the path (*i.e.*, the path constitutes the precise movement instructions) [16], some partially random movements around it (e.g., to decide how to move between two points of the target path) [43], path calculation algorithms (e.g., shortest path under different cost functions [16,34,43]), and specific maneuvers (e.g., car following, lane change, or turns) [43,44] to follow it. Nevertheless, microscopic models (where vehicles are individually modeled) such as [16,34,43,44], could allow modeling more complex and flexible behaviors by replacing the default ones. However, this is not documented in the reviewed works, and it would probably require modifying the code of the considered simulation platforms.

Most of works discard the explicit modeling of drivers, and only consider vehicle control [16,43]. Some references [43,44] include pedestrians, but only as moving objects, similar in their description to vehicles. General traffic studies are more prone to consider these issues. Their models can specify the mental state and features of people, for instance regarding their goals and ability [30], their decision-making [45], or their roles in traffic (*i.e.*, pedestrians, drivers, and passengers) [12].

Both in the case of vehicles and people, researchers lack of support to extend the provided default implementations. For instance, there is little documentation, libraries, base classes for inheritance, or schedulers for multiple tasks, in the platforms to help coding these modifications.

Our ITSML addresses these issues at several levels. It provides an extensible metamodel able to accommodate the different previous modeling approaches. The specification of environment elements and vehicles as objects with state and method interfaces is achieved with *components*. People and vehicle movements are modeled through the specification of*persons*. The case study (see Section 4.1) already includes an example of path following with maneuvers. The adoption of*agents* to represent *persons* goes further and allows describing the wide range of alternative models for people in traffic studies, from reactive to deliberative ones. Moreover, *managers* of devices are also modeled as *agents*, which facilitates representing complex distributed algorithms for ITSs including information management (e.g., goals, facts, probabilities, or uncertainty). Many issues usually considered in code are promoted to models, so researchers can define their requirements using MLs. Then, standard MDE development methodologies translate models to source code. When this translation is done, it provides documentation and reusable artifacts (*i.e.*, models and transformations) for similar implementations. Thus, it provides the basis to build improved support resources for these tasks.

The already discussed conceptual frameworks are intended to be the basis for the analysis of ITSs. They focus on the external interface of these systems (*i.e.*, behavior perceived by users), the environment, and their mutual influences. However, those references do not consider the internal organization of such systems, beyond, for instance, coding constraints imposed by simulation platforms. There are specific works focused on the architecture of ITSs. They usually present abstract models and hints for their implementation (e.g., [33,37]). Other works address low-level aspects of these systems, e.g., control or coordination mechanisms. They are usually focused on specific aspects, e.g., issues of computer vision in [41]. When dealing with such details, the use of simulations is common [7,16,42].

Works such as SUMO [16], DEUS [43], MATSim [34], the Tokyo Virtual Living Lab [44], the test platform in [35], or those reviewed in [42], are traffic simulation platforms. They include libraries to develop and run simulations, and sometimes user interfaces to specify them. Although there are descriptions of their underlying abstract theoretical models, they are black-boxes. A complete understanding of their functioning can require examining their source code when it is available. Literature has already pointed out this issue as one of the main limitations in the use of simulations for research, as it makes difficult guaranteeing the alignment between abstract models and simulations [8].

In most cases, works do not describe the development process followed. There are not explicit guidelines to reproduce modeling, design, or coding. This lack of documentation hampers validation and sharing experiences on development. Moreover, works focus on solving specific issues at a given level of abstraction, e.g., abstract model, architecture, or control of devices. They do not identify ways to integrate that research with complementary works to offer a complete solution to develop ITSs.

The current work adopts MDE to support these tasks. It provides an approach with several abstraction levels intended for different backgrounds. Researchers on ITSs work with the ITSML at the levels of CIM and PIM. Designers of simulations and programmers work mainly with the INGENIAS-ML at the levels of PIM and PSM. In both cases, experts integrate and refine models adding new information and using transformations. For the first group of tasks, this work offers modeling guidelines; for the second, INGENIAS provides a complete development methodology. The overall approach facilitates the explicit representation of all the information involved in the transition from the abstract models to code, which facilitates validation and reutilization.

6. CONCLUSIONS

This paper has presented a MDE approach for the analysis through simulations of ITSs. The complexity of these systems and the environments where they act implies high development costs. This kind of issues has been addressed in other domains through simulation. However, its application demands means to facilitate the transition from the abstract models of systems to the simulation, and its validation. Usually, researchers make these tasks manually, what is an additional source of mistakes. Adopting MDE approaches, researchers explicitly describe these steps with models and transformations. This facilitates their examination, and therefore validation and testing.

The proposed framework comprehends a ML, guidelines, and tools to develop ITS specifications and simulations. It is intended for the complete development process, focusing effort on the high-level design. Its definition reuses previous research in aspects related to ITSs. The ITSML is built over the TML [12] and the SNML [13], extending them with ITS-specific concepts. All these languages adopt the agent paradigm [11], which facilitates integrating the different perspectives.

The ITSML conceives all the elements in this domain as *components*, which are characterized by their state and a method-based interface. Flows of control and information among components use these methods. ITSs interact with their external environment using a layer of sensors and actuators, which are also *components*. Complex algorithms are modeled with*agents*. These include components for perception, reflection, and acting that manipulate information and goals.

For control activities, *sensors* perceive *events* from the environment, which they transform in internal *facts*. Their *methods*generate *notifications* from those *facts* that are communicated to other *components*, including *manager* agents. These use available information to check *goals* and decide what *tasks* to attempt. The execution of *tasks* invokes *methods*, which can generate new *notifications*. Some of these *notifications* trigger the execution of *actuator methods*, which act on the environment.

The concepts related to traffic are organized following an adapted DVE [30] model according to the structure of the TML [12]. They include common primitives such as *person*, *vehicle*, and *thing* (*i.e.*, environment element). Here, they are linked to the group of concepts corresponding to *devices* and their *manager* controllers extracted from the SNML. This last group models the interface between ITSs and their external environment.

The ITSML also provides mechanisms of inheritance and instantiation of classes in models. The first one allows flexible extensions of the ML without modifying its metamodel. The second one supports the complete definition of the components of the simulation at the model level. In this way, only algorithms need to be specified with code.

The development guidelines, along with the definition of concepts in the metamodel, help designers to build their specifications. They provide a step-by-step process to look for relevant elements in problems, and translate them to modeling primitives. The low-level design and coding follow INGENIAS [14], an existing MDE methodology for MASs. The use of agent concepts in the ITSML and the INGENIAS-ML facilitates the transition between both stages using automated transformations.

The final elements of the framework are the support tools. A model editor based on INGENME [15] supports specifying models compliant with the ITS metamodel. The code generator of the traffic framework [12] is used in the later stages of the INGENIAS process. It replaces the usual INGENIAS tools, because it provides more guidance to work jointly with models and code templates.

This framework has been used in several experiments. The case study developed a simulation of an ITS with intelligent traffic lights for the simulation platform SUMO [16]. The models specified most of the intended simulation. Only algorithms (e.g., using thresholds and timing to change the color of traffic lights, or the methods to interact with SUMO) need to be coded. As most of this code appears in templates that can be reused in code generation for different model elements, the actual coding effort is highly reduced. Changes in the simulation would only affect the model if they add new instances of available classes of entity or relationship. If new classes appear, only the templates for them need to be added before regenerating the simulation. Templates could be reused from other simulation projects. Similar findings appear in the other reported experiments, which considered simulations with extended models of decision-making, other types of ITS, and several simulation platforms.

This framework is still ongoing work with several open issues, some of them already pointed out in experiments inSection 4. First, the metamodel needs to be enriched with additional primitives. Primitives to represent *tasks* ordering or preferences are not available. Also, the specification of *vehicle* parts and their actions is not possible, for instance to represent the angle and acceleration of a vehicle turn in a junction as done in [44]. Second, certain constraints on the model, for example in the number of instances of a class, inheritance relationships, or values of attributes, cannot be expressed. The use of a constraint language such as the Object Constraint Language (OCL) [46] would reduce here the need to resort to code. Third, the development guidelines need to be extended. Now, they indicate a possible ordering of steps and leave for the metamodel definition the indications on how to identify the different modeling concepts. The identification hints should also be transferred to the process specification. Finally, additional experimentation is required to validate the framework. Current experiments are limited in number, the types of ITS modeled, and the considered simulation infrastructures.

ACKNOWLEDGMENTS

This work has been done in the context of the projects "Social Ambient Assisting Living—Methods (SociAAL)" (TIN2011-28335-C02-01) supported by the Spanish Ministry for Economy and Competitiveness. Also, we acknowledge support from the "Red Científico-Tecnológica en Ciencias de los Servicios" (TIN2011-15497-E) and the "Programa de Creación y Consolidación de Grupos de Investigación" (UCM-BSCH GR35/10-A).

AUTHOR CONTRIBUTIONS

Alberto Fernández-Isabel developed the framework for the MDE of ITS simulations, and implemented the case study. Rubén Fuentes-Fernández adapted the INGENIAS and INGENME infrastructure for the framework. All the authors contributed equally to writing the paper.

REFERENCES

1. Chourabi, H.; Nam, T.; Walker, S.; Gil-Garcia, J.R.; Mellouli, S.; Nahon, K.; Pardo, T.A.; Scholl, H.J. Understanding Smart Cities: An Integrative Framework. In Proceedings of the 45th Annual Hawaii International

Conference on System Sciences (HICSS 2012), Maui, HI, USA, 4–7 January 2012; Sprague, R.H., Jr., Ed.; IEEE: Washington, DC, USA, 2012; pp. 2289–2297.

2. Harrison, C.; Donnelly, I.A. A Theory of Smart Cities. In Proceedings of the 55th Annual Meeting of the International Society for the Systems Sciences (ISSS 2011), Hull, UK, 17–22 July 2011; pp. 1–15.

3. Tewolde, G.S. Sensor and Network Technology for Intelligent Transportation Systems. In 2012 IEEE International Conference on Electro/Information Technology (EIT 2012), Indianapolis, IN, USA, 6–8 May 2012; IEEE: Washington, DC, USA, 2012; pp. 1–7.

4. The European Parliament and the Council of the European Union. Directive 2010/40/EU of the European Parliament and of the Council of 7 July 2010, on the Framework for the Deployment of Intelligent Transport Systems in the Field of Road Transport and for Interfaces with Other Modes of Transport. Available online: http://eur-lex.europa.eu/(accessed on 15 March 2015).

5. Figueiredo, L.; Jesus, I.; Machado, J.A.T.; Ferreira, J.; de Carvalho, J.L.M. Towards the Development of Intelligent Transportation Systems. In Proceedings of the 2001 IEEE Intelligent Transportation Systems Conference (ITSC 2001), Oakland, CA, USA, 25–29 August 2001; IEEE: Washington, DC, USA, 2001; pp. 1206–1211.

6. Klein, L.A. *Sensor Technologies and Data Requirements for ITS*; Artech House Publishers: Norwood, MA, USA, 2001.

7. Pursula, M. Simulation of Traffic Systems—An Overview. *J. Geogr. Inf. Decis. Anal.* **1999**, *3*, 1–8.

8. Axtell, R.L.; Epstein, J.M. Agent-based modeling: Understanding our creations. *Bull. St. Inst.* **1994**, *9*, 28–32.

9. Bézivin, J. Model Driven Engineering: An Emerging Technical Space. In *Generative and Transformational Techniques in Software Engineering*; Lecture Notes in Computer Science; Lämmel, R., Saraiva, J., Visser, J., Eds.; Springer: Heidelberg, Germany, 2006; Volume 4143, pp. 36–64.

10. Fuentes-Fernández, R.; Hassan, S.; Pavón, J.; Galán, J.M.; López-Paredes, A. Metamodels for role-driven agent-based modelling. *Comput. Math. Organ. Theory* **2012**, *18*, 91–112.

11. Müller, J.P.; Pischel, M.; Thiel, M. Modeling Reactive Behaviour in Vertically Layered Agent Architectures. In*Intelligent Agents*; Lecture Notes in Computer Science; Wooldridge, M.J., Jennings, N.R., Eds.; Springer: Heidelberg, Germany, 1995; Volume 890, pp. 261–276.

12. Fernández-Isabel, A.; Fuentes-Fernández, R. A Model-Driven Engineering Process for Agent-Based Traffic Simulations. In Proceedings of the 5th International Conference on Simulation and Modeling Methodologies, Technologies and Applications (SIMULTECH 2015), Colmar, France, 21–23 July 2015; SCITEPRESS: Setúbal, Portugal, 2015; pp. 1–10.

13. Fuentes-Fernández, R.; Guijarro, M.; Pajares, G. A Multi-Agent System Architecture for Sensor Networks. *Sensors***2009**, *9*, 10244–10269.

14. Pavón, J.; Gómez-Sanz, J.J.; Fuentes, R. The INGENIAS Methodology and Tools. In *Agent-Oriented Methodologies*; Henderson-Sellers, B., Giorgini, P., Eds.; Idea Group Publishing: Hershey, PA, USA, 2005; Chapter IX; pp. 236–276.

15. Pavón, J.; Gómez-Sanz, J.; Paredes-López, A. The SiCoSSyS approach to SoS engineering. In Proceedings of the 6th International Conference on System of Systems Engineering (SoSE 2011), Albuquerque, NM, USA, 27–30 June 2011; IEEE: Washington, DC, USA, 2011; pp. 179–184.

16. Krajzewicz, D.; Brockfeld, E.; Mikat, J.; Ringel, J.; Rössel, C.; Tuchscheerer, W.; Wagner, P.; Wösler, R. Simulation of modern Traffic Lights Control Systems using the open source Traffic Simulation SUMO. In Proceedings of the 3rd Industrial Simulation Conference (ISC 2005), Berlin, Germany, 9–11 June 2005; Krueger, J., Lisounkin, A., Schreck, G., Eds.; EUROSIS-ETI: Ostend, Belgium, 2005; pp. 299–302.

17. Object Management Group. *OMG Meta-Object Facility (MOF) Core Specification, Version 2.4.2*; Object Management Group, 2014. Available online: http://www.omg.org/ (accessed on 15 March 2015).

18. Object Management Group. *OMG Unified Modeling Language (OMG UML), Version 2.5*; Object Management Group, 2013. Available online: http://www.omg.org/ (accessed on 15 March 2015).

19. Steinberg, D.; Budinsky, F.; Paternostro, M.; Merks, E. *EMF—Eclipse Modeling Framework*, 2nd ed.; Addison-Wesley Professional: Boston, MA, USA, 2008.

20. Lopes, D.; Hammoudi, S.; Bézivin, J.; Jouault, F. Mapping Specification in MDA: From Theory to Practice. In*Interoperability of Enterprise Software and Applications*; Konstantas, D., Bourrières, J.-P., Léonard, M., Boudjlida, N., Eds.; Springer: Heidelberg, Germany, 2006; pp. 253–264.

21. Kelly, S.; Lyytinen, K.; Rossi, M. Metaedit+: A fully configurable multi-user and multi-tool CASE and CAME environment. In *Advanced Information Systems Engineering*; Lecture Notes in Computer Science; Constantopoulos, P., Mylopoulos, J., Vassiliou, Y., Eds.; Springer: Heidelberg, Germany, 1996; Volume 1080, pp. 1–21.

22. Mens, T.; van Gorp, P. A Taxonomy of Model Transformation. *Electron. Notes Theor. Comput. Sci.* **2006**, *152*, 125–142.

23. Czarnecki, K.; Helsen, S. Feature-Based survey of model transformation approaches. *IBM Syst. J.* **2006**, *45*, 621–645.

24. Eclipse M2T. Eclipse Modeling-M2T-Home. Available online: https://www.eclipse.org/modeling/m2t/ (accessed on 15 March 2015).

25. Eclipse MMT. Eclipse Model to Model Transformation. Available online: https://www.eclipse.org/mmt/ (accessed on 15 March 2015).

26. Kleppe, A.G.; Warmer, J.; Bast, B. *MDA Explained: The Model Driven Architecture—Practice and Promise*; Addison-Wesley: Boston, MA, USA, 2003.

27. Gómez-Rodríguez, A.; Fuentes-Fernández, R.; González-Moreno, J.C.; Rodríguez-Martínez, F.J. INGENIAS with the Unified Development Process. In *Handbook on Agent-Oriented Design Processes*; Cossentino, M., Hilaire, V., Molesini, A., Seidita, V., Eds.; Springer: Heidelberg, Germany, 2014; pp. 371–405.

28. González-Moreno, J.C.; Gómez-Rodríguez, A.; Fuentes-Fernández, R.; Ramos-Valcárcel, D. INGENIAS-Scrum. In*Handbook on Agent-Oriented Design Processes*; Cossentino, M., Hilaire, V., Molesini, A., Seidita, V., Eds.; Springer: Heidelberg, Germany, 2014; pp. 219–251.

29. García-Magariño, I.; Fuentes-Fernández, R. A Technique for Defining Metamodel Translations. *IEICE Trans. Inf. Syst.***2009**, *92*, 2043–2052.

30. Amditis, A.; Pagle, K.; Joshi, S.; Bekiaris, E. Driver-Vehicle-Environment monitoring for on-board driver support systems: Lessons learned from design and implementation. *Appl. Ergon.* **2010**, *41*, 225–235.

31. Rubel, D.; Wren, J.; Clayberg, E. *The Eclipse Graphical Editing Framework (GEF)*; Addison-Wesley Professional: Boston, MA, USA, 2011.

32. Ehlert, P.A.M.; Rothkrantz, L.J.M. Microscopic traffic simulation with reactive driving agents. In Proceedings of the 2001 IEEE Intelligent Transportation Systems Conference (ITSC 2001), Oakland, CA, USA, 25–29 August 2001; IEEE: Washington, DC, USA, 2001; pp. 860–865.

33. Karpiriski, M.; Senart, A.; Cahill, V. Sensor networks for smart roads. In Proceedings of the 4th Annual IEEE International Conference on Pervasive Computing and Communications, Workshops (PerComW 2006), Pissa, Italy, 13–17 March 2006; IEEE: Washington, DC, USA, 2006; p. 310.

34. Transport Systems Planning and Transport Telematics Group (Technische Universität Berlin); Transport Planning Group (Swiss Federal Institute of Technology Zurich); Senozon Company. *MATSim, Multi-Agent Transport Simulation*; MATSim, 2015. Available online: http://www.matsim.org/ (accessed on 25 May 2015).

35. Fernández-Isabel, A.; Fuentes-Fernández, R. An Agent-Based Platform for Traffic Simulation. In Soft Computing Models in Industrial and Environmental Applications, 6th International Conference SOCO 2011; Advances in Intelligent and Soft Computing. Corchado, E., Snášel, V., Sedano, J., Hassanien, A.E., Calvo, J.L., Ślęzak, D., Eds.; Springer: Heidelberg, Germany, 2011; Volume 87, pp. 505–514.

36. Pascale, A.; Nicoli, M.; Deflorio, F.; Dalla Chiara, B.; Spagnolini, U. Wireless sensor networks for traffic management and road safety. *IET Intell. Transp. Syst.* **2012**, *6*, 67–77.

37. Sun, Z.; Bebis, G.; Miller, R. On-Road vehicle detection: A review. *IEEE Trans. Pattern Anal. Mach. Intell.* **2006**, *28*, 694–711.

38. Faouzi, N.-E.E.; Leung, H.; Kurian, A. Data fusion in intelligent transportation systems: Progress and challenges—A survey. *Inf. Fusion* **2011**, *12*, 4–10.

39. Gandhi, T.; Trivedi, M.M. Pedestrian protection systems: Issues, survey, and challenges. *IEEE Trans. Intell. Transp. Syst.* **2007**, *8*, 413–430.

40. Hancke, G.P.; de Carvalho e Silva, B.; Hancke, G.P., Jr. The Role of Advanced Sensing in Smart Cities. *Sensors* **2012**, *13*, 393–425.

41. Trivedi, M.M.; Gandhi, T.; McCall, J. Looking-In and looking-out of a vehicle: Computer-Vision-Based enhanced vehicle safety. *IEEE Trans. Intell. Transp. Syst.* **2007**, *8*, 108–120.

42. Kotusevski, G.; Hawick, K.A. A Review of Traffic Simulation Software. *Res. Lett. Inf. Math. Sci.* **2009**, *13*, 35–54.

43. Picone, M.; Amoretti, M.; Zanichelli, F. Simulating Smart Cities with DEUS. In Proceedings of the 5th International ICST Conference on Simulation Tools and Techniques (SIMUTOOLS 2012), Desenzano, Italy, 19–23 March 2012; pp. 172–177.

44. Prendinger, H.; Gajananan, K.; Zaky, A.; Fares, A.; Molenaar, R.; Urbano, D.; van Lint, H.; Gomaa, W. Tokyo Virtual Living Lab: Designing Smart Cities Based on the 3D Internet. *IEEE Internet Comput.* **2013**, *17*, 30–38.

45. Salvucci, D.D.; Boer, E.R.; Liu, A. Toward an Integrated Model of Driver Behavior in Cognitive Architecture. *Transp. Res. Rec. J. Transp. Res. Board* **2001**, *1779*, 9–16.

46. Object Management Group. *Object Constraint Language, Version 2.4*; Object Management Group, 2014. Available online:http:// www.omg.org/ (accessed on 15 March 2015).

Chapter 11

User Needs in Green ITS: Results of a Questionnaire Survey and Proposal for Green ITS Design

Wing Yan Man[1], Jing Bie [2], Bart van Arem[3]

[1]Department of Civil Engineering, Faculty of Engineering Technology, University of Twente, Drienerlolaan 5, 7522 NB Enschede, The Netherlands

[2]Centre for Transport Studies, Faculty of Engineering Technology, University of Twente, Drienerlolaan 5, 7522 NB Enschede, The Netherlands

[3]Department of Transport and Planning, Faculty of Civil Engineering and Geosciences, Delft University of Technology, 2628 CN Delft, The Netherlands

ABSTRACT

An internet survey has been conducted among drivers in the Netherlands and Japan to study their attitude towards Green ITS. The survey focuses on driving behavior and ITS experience, attitude towards environment, and Green ITS preferences. The results show that money-related information gives the most effective motivation for drivers to follow fuel efficient behavior; the least preferred information is impact on environment. Correlations are found between information preference and the driving style. The majority of the respondents intend to make frequent use of Green ITS. Based on these results, a new design for Green ITS is proposed, providing drivers with information on driving behavior, expenses on fuel consumption and environmental factors.

KEYWORDS

Green ITS; User needs; Fuel efficiency; Questionnaire survey; User interface

1 INTRODUCTION

Excessive energy consumption and greenhouse gas (GHG) emissions are making changes to our living environment that will be damaging in the long term. A prominent example is global warming, as acknowledged in the Kyoto Protocol [1]. Road traffic has been a major contributor to such changes. Now, more than ever, people are being encouraged to act upon it to create a better environment for the future. Diminishing car use would decrease GHG productions [2] but this is difficult to achieve in this modern and dynamic world [3]. A realistic solution would be to improve fuel efficiency, such as to guide our drivers to a fuel efficient driving behavior. High speeds, heavy acceleration and lack of anticipation in traffic have been identified as the main causes of excessive emission production [4]. By improving the driving skills to a more environmentally friendly style, we can reduce fuel consumption, fuel expenses as well as emission production. This benefits both the individual and the environment.

Green ITS (Green Intelligent Transport System) monitors the driver's performance in order to achieve a more fuel efficient behavior. By displaying information on driving history and CO_2 emission, the system informs drivers of their impact on the environment. Nowadays, there are a few systems available on the market that monitor driving behavior, such as Fiat eco: Drive [5] which reports on the fuel economy after the trip. In other systems, fuel consumptions are shown to the driver dynamically. To the best of our knowledge, no user feedback has been published in terms of driver response and behavioral adaptation due to Green ITS. To promote and improve these systems, the design and functioning should satisfy the desires of the user [6]. By measuring the attitude of drivers towards Green ITS, designers will be aware of the needs and preferences of drivers. Green ITS can then be designed from both a technological point of view and the preferences of the driver.

This research focuses on user needs in Green ITS. The objective is to improve current designs of Green ITS by considering factors that motivate drivers to change their behavior. Based on the results of an internet survey, this paper will elaborate on user attitude and preferences towards Green ITS among Dutch and Japanese drivers. A new design of Green ITS is then proposed which motivates drivers to follow environmentally friendly driving behavior.

2 BACKGROUND

To achieve sustainable mobility, various measures are being implemented to reduce GHG emissions around the globe. Traditional measures include increasing capacity of the road network (such as constructing extra lanes) and imposing fuel tax for transport [7]. The car industry has been active in introducing electric/hybrid cars and information technology systems that support the driving task [8]. For reducing GHG emission, researchers have been focusing on engine technologies, policy measures, changes to the infrastructure, modal shift, transport management, and driving behavioral changes.

In this paper we focus on the driving behavioral changes. For most drivers it is not easy to change their current driving behavior by adopting new skills. To help the driver in changing his (or her) behavior, in-car Green ITS systems and services are being developed. These systems aim at achieving a fuel efficient behavior. They provide the driver with information on his driving behavior: the driver is made known of his performance by several indicators, displayed during and/or after the trip. Examples of the indicators include characteristics of fuel consumption and speed profile. More advanced systems collect driver behavior data from previous trips and compare them with the optimal fuel efficient behavior. Personal advices can then be provided to the driver to achieve the optimal fuel efficiency. It is also possible that these advices are given in real time (i.e. during the trip, while the driver is driving). Such a system monitors the current driving behavior and intervenes if inefficient behavior is being practiced. The driver receives direct feedbacks on his behavior (progression) and advices on changing his behavior for better fuel efficiency.

3 QUESTIONNAIRE SURVEY

For a successful design of in-car Green ITS systems and services, user needs and acceptance are essential. We have deployed a questionnaire survey to investigate drivers' opinions towards Green ITS. The purpose of this survey is to assess the following aspects:

- Type of preferred information for motivating the driver to change into fuel efficient behavior;
- Influence of driver's personal characteristics on the type of preferred information;
- Attitude towards Green ITS.

The currently available in-car technologies are mostly designed from a technological point of view. With the knowledge of driver attitude and preferences, we can then improve the current designs of Green ITS.

3.1 Questionnaire Design

The questionnaire survey "Green light for a better environment" has been deployed and it consists of questions on the following four subjects:

- General information;
- Driving behavior & ITS experience;
- Attitude towards environment;
- Attitude towards Green ITS.

The survey first provides the respondent with a brief introduction on Green ITS. This introduction does not include the detailed attributes of Green ITS (such as the possibility of different interfaces, sounds and lights), in order to avoid confusion and pre-selection bias.

3.1.1 General Information

This part consists of questions on the personal characteristics of the respondent. Only respondents possessing a driving license are asked to participate in the survey. The questions are related to age, occupation, type of driving licenses, vehicle ownership, and vehicle usage (frequency and vehicle features). With this information drivers can be categorized. Factors such as driving experience and personality can also be easily retrieved from the dataset.

3.1.2 Driving Behavior & ITS Experience

Questions in this part are related to average mileage, driving frequency, trip purposes, route choice, and familiarity with ITS. By investigating the respondent's driving behavior we can identify which driver group yields the highest impact on the environment. This driver group needs special support from Green ITS. Also, driver experience with ITS is studied to assess driver familiarity with ITS technologies, which influences driver adaptability towards Green ITS. This is an important factor in the design of Green ITS.

3.1.3 Attitude Towards Environment

Drivers are asked to rank several types of information in terms of the likelihood to trigger a behavioral change on them. Other questions relate to the attitude towards environment in general and the self image of the respondent on environmental friendliness. By measuring driver attitude towards the environment we can identify the type of drivers who have a positive attitude and are therefore more willing to change their behavior for a better environment. Drivers who are not impressed by the environmental issues need other stimulus for changing their behavior.

3.1.4 Attitude Towards Green ITS

To design a Green ITS system that satisfies the desires of the driver, we need to measure the driver's attitude. Many drivers are not yet familiar with in-car Green ITS, because it is still a novelty. To measure their attitude towards such systems, examples of in-car Green ITS are provided to give the respondents a reasonable impression. Respondents are then asked which type of information is relevant to help them in adapting their behavior. Four categories of information are considered:

- general information such as fuel consumption, costs and CO_2emission;
- advice on driving behavior;
- analysis of previous driving behavior;
- progress in driving behavior.

Besides the type of information, the timing of information provision is also studied.

3.2 Conducting the Survey

Internet survey has been chosen as survey method. It has the advantages of easy accessibility, low costs, direct collection of digital data, and convenience in data processing. Internet surveys provide an effective way of reaching the drivers and collecting their opinions. In this study we only consider drivers from the Netherlands and Japan. As of June 2010, 88.3% of the whole population in the Netherlands are internet users. The same statistic is 78.2% for Japan [9]. We expect the ratio of internet users among Dutch and Japanese drivers is around or even higher than these percentages. Therefore internet surveys do not suffer greatly from the selection bias of excluding non-internet users.

The survey is set up in both the Dutch and Japanese languages. Limeservice (www.limeservice.com) has been chosen as the survey interface owing to its ability to edit both Latin and Japanese characters. The final web survey is hosted on the following web pages:

- ***Dutch version:*** http://wingyan.limeask.com/index.php?sid=56984&lang=nl
- ***Japanese version:*** http://wingyan.limeask.com/index.php?sid=97472&lang=ja

The survey was conducted in the period between December 2008 and February 2009. The respondents were informed of the survey by an email which included information on the research and a link to the survey webpage. Respondents in the Netherlands were reached through a personal network. They were also asked to forward the questionnaire among their social networks. To reach the Japanese respondents, the survey was spread over companies and universities. This snowball method turns out to be quite effective, due to the efficiency in email services and the flexibility of a web survey.

4 DATA ANALYSIS

In total 100 respondents from the Netherlands and 98 respondents from Japan filled in the survey. Table 1 shows the general characteristics of the respondents. For both countries the majority of the respondents are between 18 and 39 of age, and students in profession. There is no professional driver respondent from Japan. The variation in profession is higher in the Netherlands than in Japan.

Table 1. Characteristics of respondents

Characteristics		Dutch	Japanese
Age	18-24	48%	40.6%
	25-39	36%	38.5%
	40-64	13%	19.8%
	65 and older	3%	1.0%
Profession	Student	39%	53.1%
	Office worker	28%	30.2%
	Professional driver	5%	0.0%
	Other	28%	16.7%
Car possession	None	32%	44.8%
	1	38%	45.8%
	2-3	28%	9.4%
	More than 3	2%	0.0%
Annual average mileage	0 - 1.000 km	11%	41.7%
	1.000 – 5.000 km	19%	29.2%
	5.000 - 10.000 km	13%	14.6%
	10.000 - 20.000 km	29%	12.5%
	20.000 - 30.000 km	13%	1.0%
	More than 30.000 km	15%	1.0%
Primary trip purpose	Commuting	41%	15.6%
	Work related	13%	3.1%
	Social & Recreational	45%	56.3%
	Shopping	1%	25.0%

Between the respondents from these two countries, significant differences are observed in terms of car possession, annual average mileage and primary trip purpose. Some Dutch drivers possess more than one car, while most of the Japanese drivers surveyed have one car or no car. Dutch respondents have a higher average mileage, make more frequent use of the car, and are more likely to commute by car. On the contrary, Japanese respondents have a lower mileage, use the car occasionally and mainly for private situations.

These differences are attributed to the different demographics in these two countries. Urban population density is higher in Japan than in the Netherlands. Ridership of public transport is also much higher in Japan. Using car to commute is far less popular in Japan than in the Netherlands. Japanese drivers often use the car for more private purposes, such as social and recreational activities.

The differences are also partially attributed to the unbalanced representation of the population within the respondents. As noted above, young people and students make out a big proportion of the respondents. The group of the respondents surveyed in this study is therefore an unrepresentative sample of the society. This bias will be taken into account in our analysis below.

4.1 Information Priority

Instead of categorizing drivers by their personal characteristics and then examining per group the preferred information, we have done the analysis in reverse. In the survey the respondents are asked to select the most relevant information for them to receive in order to change their behavior; they had to rank four types of information by priority. The results provide the reference point in our analysis: this priority information defines the respondent's character and attitude towards environment. By correlating this character to his driving style and experience with ITS, suitable Green ITS based on the preferred information can be designed.

Table 2 shows the frequencies of the four types of information being ranked with the highest priority. For both countries, money related information is the most important to drivers. The second most preferable information is about fuel consumption; this information is more preferable for Japanese drivers than for Dutch drivers. A remarkable difference is the importance of CO_2 emission between the two countries.

Table 2. Distributions of priority information

Priority information	Dutch	Japanese
Fuel consumption	17%	34.4%
Money saving	64%	56.3%
CO_2 emission	11%	2.1%
Contribution to environment	8%	7.3%

4.1.1 Personal Character

Several key factors have been chosen to uncover correlations between the preferred information and the character of the driver. To define the personal character of the driver, three key factors have been analyzed: age, profession and car possession. Age and profession tell something about the driver's daily life. Generally, office workers drive or commute more by car than students, which makes them a more important target for Green ITS. Car possession indicates the driver's need of a car and/or desire to drive. These factors are then compared with the preference of information.

The survey results show that Dutch respondents between the age 18 and 24, being students, and without car possession, are much more interested in money related information than other driver groups. The fact that office workers are forced to use the car and usually can declare their expenses to their employer explains the importance of fuel consumption instead of expenses. For the Japanese respondents, no correlation between choice of priority information and personal character has been identified. Each driver group (by age, profession or car possession) exhibits a distribution of priority information similar to the overall distribution in Table 2.

4.1.2 Driving Style

Driving style characterizes the driver's driving behavior in daily life. The annual mileage and driving frequency provides information on car use. Drivers with a higher mileage produce more emissions. Primary trip purpose reveals the driver's familiarity with the routes during an ordinary trip. Commuting is often via the same route, meaning that the driver is familiar with the route and more likely to drive in a smooth way. This leads to less emission compared to recreational trips, where the driver is unfamiliar with the routes and has to learn about the conditions on these routes.

The survey results show that Dutch respondents with an average to low annual mileage (1,000 km to 10,000 km) and less frequent use of car (1 to 8 times per month) are much more interested in money savings than other driver groups. No major difference has been identified between trip purposes. For the Japanese respondents, it appears that annual mileage and driving frequency have very little influence on the choice of priority information. However, respondents with their primary trip purpose as private (social recreational and shopping) are slightly less interested in money savings than other driver groups.

4.1.3 Experience with ITS

Drivers' previous experience and familiarity with ITS systems and services make it easier for them to adapt to Green ITS, if the Green ITS is designed in a comprehensible way and in accordance with other ITS systems. In this survey, respondents are to rate their familiarity with cruise control, adaptive cruise control, lane departure warning, navigation systems, and intelligent information systems.

The survey results show a striking difference between the two countries in terms of the respondent's familiarity with ITS systems. Japanese respondents appear to be significantly less familiar with current ITS systems than the Dutch respondents, possibly due to the large amount of students in the respondent group.

Nonetheless, cruise control and navigation systems are widely used and perceived as familiar in both countries. Therefore a comprehensible new system should have the same level of adaptability as these two systems. For example, some interface aspects of these systems can be borrowed to maintain familiarity with the new system.

However, familiarity with ITS is shown to be not strongly related to the choice of priority information. Among drivers who are experienced with a particular ITS system, similar ratios in information preference are observed as those in Table 2.

4.1.4 Attitude Towards Environment

Environmentally friendly behavior concerns actions that people take to improve the environment, such as recycling or low energy use. Most respondents hesitate to call themselves very environmentally friendly. The majority of respondents reflect on themselves as having an average personality towards environmental friendliness. Dutch respondents with CO_2 emission and contribution to environment as their priority information do think that they are more environmentally friendly. On the contrary, Japanese respondents who are interested in information on their contribution to environment mostly refer to themselves as not environmentally friendly.

The respondents are also asked to reflect on their own fuel consumption behavior during driving. Respondents who choose contribution to environment as their priority information generally consider their driving behavior as not very fuel efficient, indicating their need to acquire such information. On the other hand, respondents who consider themselves as fuel efficient mainly choose fuel consumption as their priority information, signifying their desire to further reduce their fuel consumption.

Another question in the survey is about the influence of fuel price on car use. The intention of raising fuel price is to reduce fuel consumption. It is expected that respondents with fuel consumption and money savings as their priority information would be highly affected by fuel price. However, this is not the case in the survey. For the Japanese respondents, the effect of raising fuel price is independent of the driver's priority information; for each of the four respondent groups (as defined by Table 2), about half of the respondents report fuel price as having an influence while the other half state no influence. For the Dutch respondents, raising fuel price has limited influence on their behavior. For those with money savings as their priority information, the majority of them actually claim that fuel price has no influence at all on their behavior. A possible explanation is that these drivers are dependent on their car for commuting and cannot reduce their frequency of car use even under high fuel prices.

4.2 Relevant Information for Changing Driving Behavior

In the survey respondents are asked to rate several types of Green ITS information in terms of relevance. A high relevancy is interpreted as a high willingness to use Green ITS and to adapt one's driving behavior according to the received information. Information with the highest relevancy is the most preferable for the drivers to receive. Next to the types of information, respondents are also asked to indicate their preferred timing of receiving the information (discussed in section 4.3).

4.2.1 Monetary Impact

The majority of drivers chose money related information as the most relevant motive for them to change their behavior. These drivers also include the most frequent drivers, i.e. those who produce the most emissions and with whom the most impact of Green ITS can be achieved. The information can be directly or indirectly related to money, including fuel consumption, money converted from fuel consumption, and the amount of money saved with fuel efficient behavior (Figure 1). This type of information is much more relevant for Japanese drivers than for Dutch drivers, although more Dutch respondents have chosen money related information as their priority information than Japanese respondents.

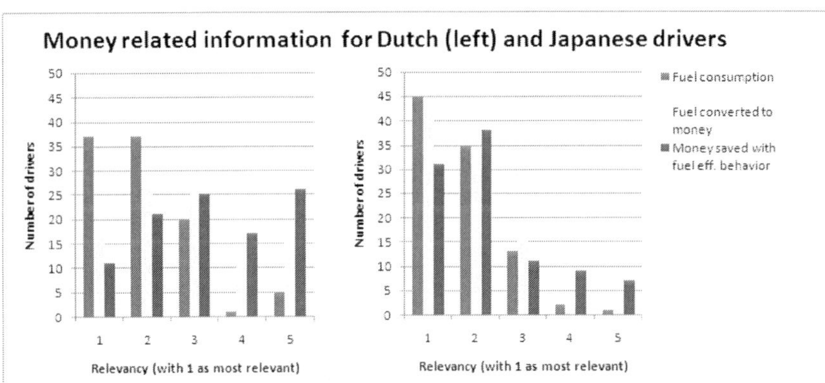

Figure 1. Level of relevance for money related information: Dutch drivers (*left*) and Japanese drivers (*right*)

4.2.2 Environmental Impact

As for the types of environment related information, the indicator for fuel efficiency is the most popular among both Dutch and Japanese drivers (Figure 2). For Dutch drivers, information on CO_2 emission and contribution to environment are in general irrelevant, although the number of drivers who do indeed rate such information as relevant is not negligible. However, Japanese drivers generally find environment related information as relevant and are willing to receive such information. A possible reason is curiosity, i.e. the eagerness among Japanese drivers to use new technology and to receive all kinds of information.

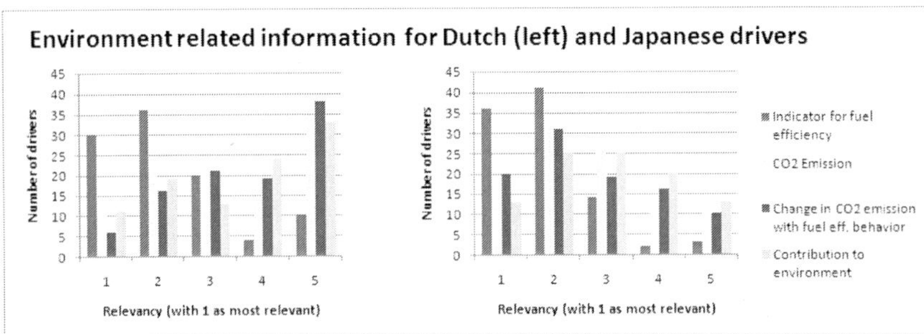

Figure 2. Level of relevance for environment related information: Dutch drivers (*left*) and Japanese drivers (*right*)

4.2.3 Personal Driving Behavior

Five elements of personal driving behavior are considered: speed, gear shift, acceleration, deceleration, and route choice. Information related to such behavior includes the advice on optimal behavior, the en route reporting of driving progress, and the post trip reporting of driving history. Both Dutch and Japanese drivers find advices on their behavior very relevant (Figure 3). Moreover, Japanese drivers show an equal interest in receiving information on their driving progress and history, whereas Dutch drivers find such information less interesting than advices. In terms of the behavioral elements, Dutch drivers prefer to receive information on speed and gear shift than on acceleration and deceleration, while Japanese drivers are indifferent.

4.3 Timing of Information

In the design of Green ITS, timing of the information supply is very important. Wrong timing of information can reduce the impact on driving behavior. An example is when advice on behavior is only given after trips. After trips drivers usually do not care about their behavior anymore and for the next trip they are more likely to have forgotten the advice.

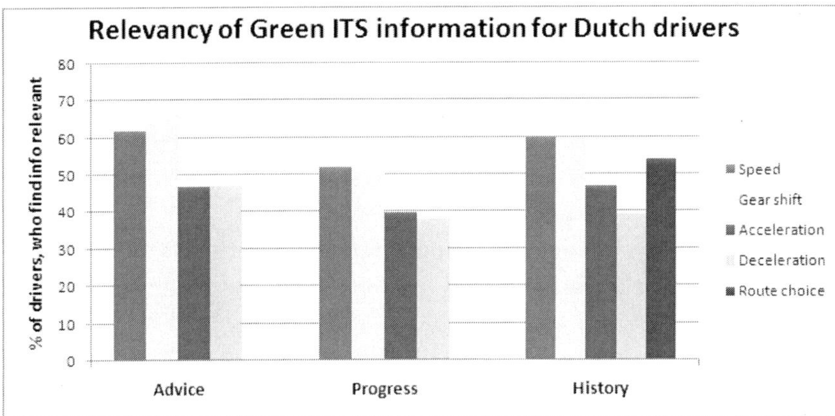

Figure 3. Level of relevance for different types of Green ITS information: Dutch drivers (*left*) and Japanese drivers (*right*)

In the survey respondents are presented with control-related advices and money-related information. They are then asked for their preferences on the timing of these advices and information. In terms of control-related advices, both Dutch and Japanese drivers prefer to receive such information during the trip; this preference is especially strong among Dutch drivers. In terms of money-related information (Figure 4), Dutch drivers prefer to receive information on fuel consumption during the trip and information on money savings after the trip. Japanese drivers, however, prefer to receive both types of information during the trip.

5 PROPOSAL FOR GREEN ITS DESIGN

Based on results of the internet survey, improvements can be made to current Green ITS systems to better accommodate drivers' preferences. Accordingly, a new Green ITS design is proposed here.

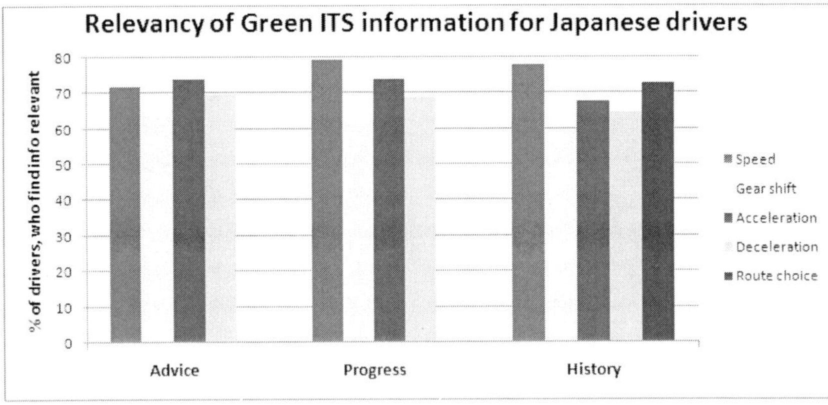

Figure 4. Preferred timing of the money related information: Dutch drivers (*left*) and Japanese drivers (*right*)

5.1 Data Flow

History, progress and advice on driver behavior are based on the personal driving behavior, i.e. the factors of speed, gear shift, and ac-/deceleration. During the trip, data of these factors are collected and then compared to the optimal fuel efficient behavior (e.g. optimal speed, optimal gear shift moment) in order to determine the advice to be given to the driver. Once the driver is following the advice, the difference due to this behavioral change is calculated.

Besides information on personal driving behavior, other types of information are also available to the driver. Money related information includes fuel consumption in liters per kilometer, fuel consumption calculated in monetary values, and the amount of money savings in fuel consumption (compared to the "previous" behavior). Environment related information includes the amount of CO_2 emission and the reduction of CO_2emission.

5.2 Type of Information

The use of Green ITS should not affect the driver's safety. To avoid stress, distraction and confusion during driving, drivers can choose to only display the essential information directly related to their driving behavior such as progress of the personal driving behavior. Other types of information can be displayed before and/or after the trip. Information that has direct effects on the driving behavior is advice for fuel efficient behavior. Unlike information about driving progress and history, advice can be directly adapted by the driver. The driver has the choice to display advice for speed, gear shifts, and ac-/deceleration.

For money related information, it is suggested to only display fuel consumption during the trip. Because of changing fuel prices, monetary values are not always accurate and are better displayed before or after the trip (with the driver being able to specify the fuel price). All environment-related information can be made available after the trip.

5.3 User Interface

Figure 5 gives an impression on the proposed Green ITS system. During the trip the information displayed should be easily understandable and without any confusion or distraction. An on board unit is attached to the dashboard with its screen showing information on gear shift, and ac-/deceleration. The screen will gradually turn red when inefficient behavior is detected, such as excessive pedal pressure and gear shift at high revolutions. The deeper the color, the more inefficient the behavior is. With fuel efficient behavior the display turns into green again. The bar in the corner displays how much fuel is left. The factors of speed, gear shift, and ac-/deceleration are displayed in line charts on screen, showing both the actual behavior (the red curve) and the optimal behavior (the green curve).

Graphs displayed after the trip provide overviews on the differences between the actual fuel consumption and the fuel consumption under fuel efficient behavior (in both volume and money). This tells the driver how much excessive fuel has been consumed and how much money could have been saved with fuel efficient behavior. The same concept applies to environment related information, i.e. differences between the actual CO_2 emission and the emission under optimal behavior.

Figure 5. Conceptual sketch of the proposed Green ITS system

6 CONCLUSIONS AND DISCUSSIONS

Through an internet survey we studied the attitude towards Green ITS among drivers in the Netherlands and Japan. It has been shown that the biggest motivation for drivers to follow fuel efficient behavior is the potential money savings. Green ITS should strive to provide drivers with information on how much money they would save. Next to that, drivers are also interested in information on fuel consumption during their trips. The least preferred information is actually about the impact on environment, such as CO_2emission. The cause of this disinterest could be that drivers do not see a direct relationship between their car use and its impact on the environment. In order to make these drivers aware of their impact on environment and to motivate them to adapt to environmentally friendly behavior, we need to help the driver establish a direct, understandable and imaginable relationship between their behavior and the environmental impact. An example could be to make a comparison between emission and the reduction of trees.

Personal characteristics, experience with ITS, and attitude towards environment have been shown to cast only a minor influence on the driver's preference of information in Green ITS. Correlations, however, have been found between the preference of information and the driving style. Overall, the majority of respondents consider information provided in Green ITS as relevant and they are willing to use such systems. This positive attitude is expected to contribute to high acceptance and frequent use of Green ITS. However, the limited scope of this study should be noted in future user preference studies where more representative respondent sampling should be made. It should also be noted that drivers' actual preference and behavior (revealed preference) may turn out to be different than what they have stated in the questionnaire survey (stated preference).

Green ITS is a relatively new concept. Very limited deployment experiences have been reported in literature. For future implementation and promotion of Green ITS, information on driver experiences should be gathered. By analyzing these experiences, likes and dislikes of users can be identified, which will make room for further improvement. These improvements can be based on the type of information supply, the timing of information supply and the display of information. Different types of information or modes of display are expected to have different effects on the driving behavior.

In the design of Green ITS, safety, distraction and other psychological influences on the driver are of great importance. In the proposed design, the system provides information on driving behavior, expenses on fuel consumption and environmental factors. During the trip, advice and fuel consumption will be displayed with simple indications on an on-board screen. After the trip other information will be presented with trends, numbers and texts.

ACKNOWLEDGMENTS

The research described in this paper was conducted during the first author's bachelor assignment at Toyota InfoTechnology Center Co., Ltd. The authors would like to thank Sachi Mizobuchi and the Department of Research and Development for their enthusiasm and contribution to this research. The authors also acknowledge the constructive comments of an anonymous referee.

REFERENCES

1. UNFCCC, Kyoto Protocol. Available online: http://unfccc.int/kyoto_protocol/items/2830.php (1997)

2. Logghe, S., Van Herbruggen, B., Van Zeebroeck, B.: Emissions of Road Traffic in Belgium. Transport & Mobility Leuven (2006)
3. Fergusson, M., Smokers, R., Passier, G., ten Brink, P., Watkins, E., Valsecchi, C., Hensema, A.: Possible regulatory approaches to reducing CO2 emissions from cars. European Commission, Brussels (2007)
4. Bell, M.C.: Environmental factors in intelligent transport systems. IEE Proceedings: Intelligent Transport Systems **153**, 113–128 (2006)
5. FIAT, eco:Drive. Available online: http://www.fiat.co.uk/ecodrive/ (2009)
6. van Driel, C.J.G.: Driver Support in Congestions. TRAIL Research School, Delft (2007)
7. van Breugel, P.B., Baum, A., Calovi, L., Hackman, M., de Gier, C.W., Juneholm, M., Klaeboe, R., Pucher, E., Kampfer, A., Vinot, J.: Examples of air quality measures near roads within Europe: National measures of the international CEDR air quality group. Rijkswaterstaat, Delft (2005)
8. van Maarseveen, M.F.A.M., van der Voort, M.C., van Driel, C.J.G.: Qualitative Survey on Fuel Economy Devices. University of Twente, Enschede (2002)
9. de Argaez, E., Internet World Stats. Available online: http://www.internetworldstats.com/ (2011)

Chapter 12

Incident Detection in Heavy Traffics in Tunnels by the Interlayer Feedback Algorithm

Shunsuke Kamijo , Kaichi Fujimura

Institute of Industrial Science, The University of Tokyo, 4-6-1 Komaba

ABSTRACT

In this paper, we developed video surveillance system to detect incidents from heavy traffics in tunnels. Incidents in tunnels may cause additional accidents or fires that may result in fatal disaster. Therefore, it is important to detect incidents as soon as possible to manage the traffics inside the tunnels by the officers. Generally, video images in the tunnels suffer from heavy occlusion due to the low position of camera settings. In particular, the problem of heavy occlusions would be more serious in the urban tunnels due to their heavy traffics. We developed a tracking algorithm to segment vehicles and estimate the precise vehicle trajectories against the heavy occlusions. Utilizing this tracking algorithm, dedicated algorithm to detect incidents from the traffic images was developed. For the experiments, video streams of three cameras for 6 months were investigated, and 32 incidents were examined to evaluate the developed algorithm.

KEYWORDS

Incident detection; Tracking; Occlusion; Segmentation; Semantic hierarchy

1 INTRODUCTION

There are a lot of previous works on the vehicle tracking. Historical works of the vehicle tracking are as follows. A work by H. Kolling and H. Nagel [1] is widely known as successful by using vehicle shape model consisting of horizontal and vertical edges. Peterfreund [2] developed boundary extraction method by Snakes [3]' for vehicle tracking. Smith [4] and Grimson [5] employed statistical analysis and clustering of optical flows for the tracking problem. Although those works do not deal with occlusion problem, some previous works deal with the problems. Leuck [6] and Gardner [7] employed 3D models of vehicle shape and developed a dedicated pattern matching algorithm to vehicle images. However, since this method requires various kinds of vehicle models, it would be difficult to be applied to a scene containing various kinds of vehicle traveling into various directions. Malik [8] tried to resolve occlusion problem by a dedicated reasoning method. Although, those methods achieved good performance in a sparse traffic with simple motion, they cannot deal with complicated traffics.

In recent years, the tracking algorithm has been restlessly improved [9–13]. Among them, Kanhere [10] tried to solve the occlusion problem at low camera angle by employing a 3-D perspective mapping from the scene to the image along with a plumb line projection. Zhang [12] applied multilayer framework to the occlusion problem. In this work, occlusion is detected by evaluating the compactness ratio and interior distance ratio of vehicles, and the detected occlusion is handled by removing a "cutting region" of the occluded vehicles. Ambai [13] developed the tracking method against occlusions consisting of feature point detection and clustering feature points. Gradient based feature detector was used for the feature point detection, and graph cut method was used for dividing the occluding objects.

A lot of researches on incident detection have been performed. Cuchira [14] applied traffic rule-based reasoning to images for simple traffic conditions on a single-lane straight road. Jung [15] tracked vehicle from traffic images for precise traffic monitoring. Zhang [16], Srinivasan [17], and

Byun [18] employed Bayesian Network and Neural Network in order to distinguish incident from ordinary traffics. All the above algorithms do not consider about the effect of occlusion. Thus, they have some difficulties in being applied to complicated traffic situation such as congestion. Thomas [19] and Zheng [20] tried find outliers in the traffic flow statistics for the incident detection. Although their methods were successful, the methods require longer time for the detection than video analysis. Shehata [21], Versavel [22], and Austin [23] employed video surveillance for the incident detection. Although their methods were successful, they are focusing on relatively easier cases such as the detection of vehicles stalling at prohibited areas.

Incident detection in the heavy traffic is quite challenging because of heavy occlusions. The incident detections in the tunnel become more difficult due to the low angle condition of the camera settings. In this paper, we resolved the heavy occlusion problems by the interlayer feedback strategy, and developed a successful algorithm to detect incidents even in the congested traffic in the tunnel. The algorithm was examined using video images recorded in the tunnel at the Tokyo Metropolitan Expressway.

2 ALGORITHM HIERARCHY FOR SCENE UNDERSTANDING

Semantic Hierarchy is one of popular idea to construct cognitive systems. In this section, an algorithm hierarchy for traffic scene understanding is discussed in detail. In particular, an innovative idea of interlayer feedback algorithm was developed in this paper, and the idea will be described in the last part of this section. Although the discussions exemplify traffic scene understanding, the idea would be useful for cognitive systems in general.

2.1 Overview of Semantic Hierarchy

Generally, algorithm hierarchy for cognitive systems can be separated into two groups of layers, Signal Processing Layer and Semantics Layer (Figure 1). Signal Processing Layer is responsible for transforming the electronic signal such as image data into designated data having significance in the real-world such as vehicle trajectories. Semantics Layer is responsible for transforming data obtained from Signal Processing Layer into semantic symbols verbally recognizable by human.

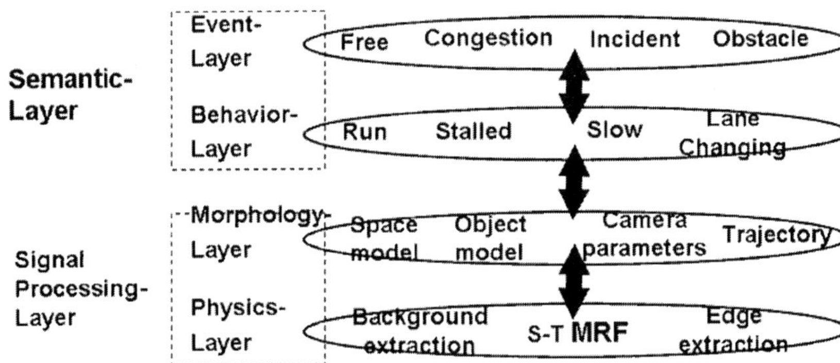

Figure 1. Algorithm hierarchy for the traffic surveillance

Physics Layer in Signal Processing Layer should contain algorithms that refer purely intrinsic information within images themselves. In other words, the algorithms in Physics Layer should not refer to any information such as forms of vehicles and human, coordinate geometry between the image and the real world, and topology of vehicle lanes. Typically, the methods for background image acquisition or edge extraction are considered as Physics Layer algorithms. Thus, such the algorithms can be generally applied to a variety of objects and scenes. On the other hand, the algorithms in Morphology Layer should refer to external information that cannot be obtained from images themselves. Models for vehicle shapes or human bodies, real world geometry, camera parameters, and vehicle lane topology are examples for such the external information.

Previous works on object tracking does not usually distinguish Physics Layer and Morphology Layer. However, it is important to clarify the difference of the two layers and to combine them appropriately, in order to maintain both benefits of the generality of Physics Layer and the respectiveness of Morphology Layer. In this paper, the idea of interlayer feedback will be discussed in detail as can be read in Section 3.

In Behavior Layer of Semantics Layer, vehicle trajectories obtained from Signal Processing Layer are transformed into verbally significant symbols that represent each vehicle's behavior such as stalling, slow, and changing lanes. Finally, algorithms in Event Layer determine the most preferable event to represent the traffic scene by combining behaviors of vehicles obtained from Behavior Layer.

As an example, when the system detected a stalling vehicle and the following vehicles changing lanes to avoid the stalled vehicle frequently, it should be determined as a scene of an incident. On the other hand, even when the system detected a stalled vehicle, it should be determined as congestion if the following vehicles do not move to the same way.

2.2 The S-T MRF Model as a Physics Layer Algorithm

The Spatio-Temporal Markov Random Field model [24, 25] was developed in order to segment spatio-temporal images by referring to purely intrinsic information within image sequences. It divides an image into blocks as a group of pixels, and optimizes labeling of such blocks by referring to texture and labeling correlations among them, in combination with their motion vectors. Combined with a stochastic relaxation method, the S-T MRF optimizes object boundaries precisely against occlusions.

For example, Object-map representing segmentation result map of object ID distribution will be obtained as shown in Figure 2. This result is obtained by applying the S-T MRF model to the image from very low-angle at merging traffic on an expressway. In the tracking images of Figure 2a, bounding boxes overlaps each other. However, vehicles were segmented correctly and the overlapped rectangles are just a problem of visualization as appeared in the Object-map in Figure 2b.

Since the S-T MRF model was defined as an algorithm in Physics Layer, it is not able to segment occluded objects moving in similar motions along the traveling. Interlayer feedback strategy would be effective to solve such the occlusion problems with similar motions as described in the following section.

3 INTERLAYER FEEDBACK ALGORITHM

Occlusion problems frequently occur in heavy traffic in the tunnel, and the degrees of occlusions would be serious due to the low angle condition of the camera installation. In addition, occluded vehicles travel to the same direction in the similar velocities in tunnel traffic as shown in Figure 3a. The S-T MRF model is not able to segment the occluded vehicles with similar motions.

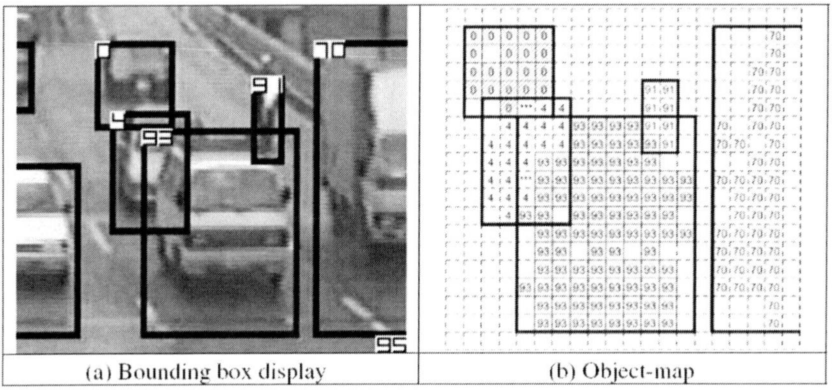

Figure 2. Results from the S-T MRF model

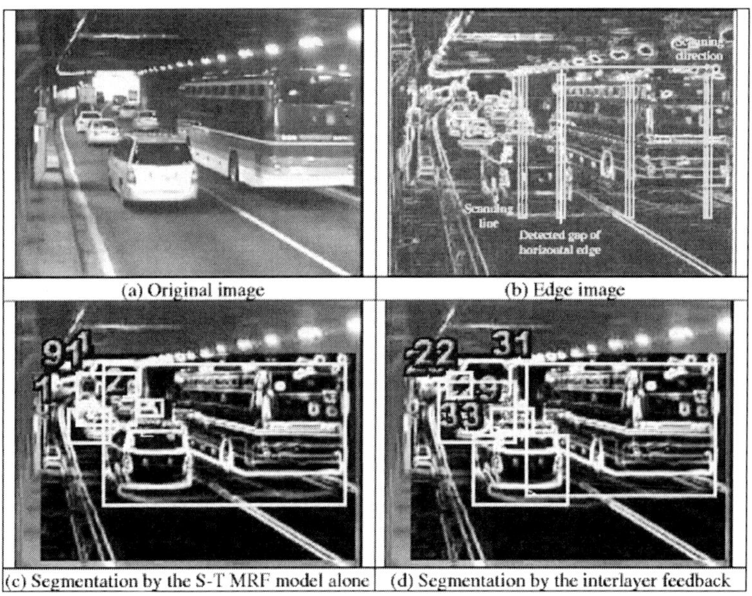

Figure 3. Overview of the interlayer feedback algorithm

In such cases, algorithms in Morphology Layer should segment the vehicles which were not segmented by the S-T MRF model. For that purpose, we developed a dedicated algorithm to segment occluding vehicles by referring to horizontal edge patterns of vehicles. Vehicle images are rich in horizontal edges, and the patterns are different among the variations of vehicles as sedan, wagon, trucks and buses as shown in Figure 3b. Therefore, occluding vehicles can be segmented by finding the gaps of horizontal edge patterns. An edge image should be obtained from an original image to find the gap. This is binary edge image defined as a binarized image of illumination invariant edge image used in S-T MRF model [25]. The coordination among S-T MRF model, edge extraction, and this algorithm (Interface-split) is represented in Figure 4. The Interface-split algorithm is applied to the rectangle judged that two vehicles exist astride two lanes. A description of the algorithm is as follows.

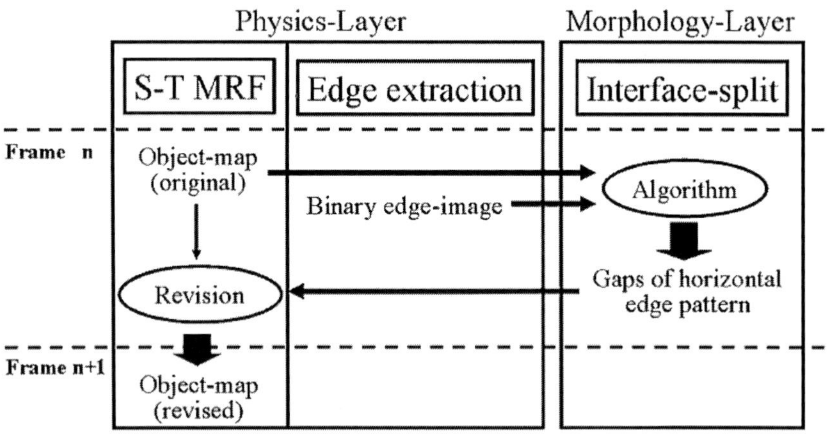

Figure 4. Diagram of interface-split

[Interface-split: detecting gap of horizontal edge]

[Step 1]

The scanning line is prepared in a vertical direction of a rectangular image, and the presence of the edge in each block on the scanning line is inspected.

[Step 2]

The one-dimensional pattern is generated on the scanning line. This pattern has a binary value corresponding to the existence of edge.

[Step 3]

The scanning line is slid to the horizontal direction of the rectangle. A line which has the maximum amount of change in the patterns between the own line and the neighbor line is detected as a splitting line. Hamming distance is used as an index of the amount of change for the line.

[Step 4-a]

If the splitting line exists, and if the amount of change for the splitting line exceeds a threshold, object ID on the Object-map is changed based on the splitting line. On the other hand, it is judged that only one vehicle (such as heavy vehicle) exists in the rectangle extracted by the ST-MRF model, and the Object-map is not changed.

[Step 5]

In the next frame, the ST-MRF model revised the Object-map based on the result of step 4.

[End of the algorithm]

In this paper, our original method for illumination invariant preprocessing was applied to the original image to cancel the effect of head lights as Figure 3b, c and d and tracking results in the following sections. The existence of an edge is examined by each block defined in the S-T MRF model, in order to reduce the time and maintain the robustness of the gap finding process.

It is remarkable that after the S-T MRF model is informed segmentation results from the algorithm in Morphology Layer, the S-T MRF model is able to segment the vehicles in the following image frames. Generally, the algorithm in Morphology Layer is not able to perform segmentation and tracking so robustly by itself. Since the S-T MRF model accepts any algorithms in Morphology Layer, the strategy of interlayer feedback is promising to apply to various kinds of objects.

4 OCCLUSION ROBUST ESTIMATION OF VEHICLE TRAJECTORIES

For the incident detection in the tunnel, evaluations of the slow down behaviors and lane changing behaviors of vehicles are necessary. Spatio-temporal estimations of trajectories enable evaluation of vehicle behaviors and scene understanding as incident detection.

In images of low angle cameras in tunnel, a majority of vehicles are not occluded when they come into the view beneath the camera. Therefore, grounding position of the vehicle tail can be estimated from the vehicle segmentation results at the moment when they travel beneath the camera. However, as the vehicles travel toward the downstream in the camera view, tail of the preceding vehicle should be occluded by the following vehicle to cause the failure in estimating grounding position of the preceding vehicle. In this section, a dedicated method to estimate the grounding position of the vehicle tail was developed as described below.

Cameras were originally installed at about 2.5 m high in the tunnel and their optical axes are horizontal. We used them as the original settings. In Figure 5, y represents vertical coordinate in the image coordinate measured from the vanishing point, and it should be transformed into real world coordinate for the estimation of vehicle velocity in the real world. When the tail of the preceding vehicle is occluded by the following vehicle, y_{grnd} of the preceding vehicle cannot be observed directly. Therefore, y_{grnd} can be estimated from y_{top} and $h_{app}(y_{grnd})$ where y_{top} represents position of the vehicle roof in the image coordinate, and $h_{app}(y_{grnd})$ represents appearance height of the vehicle in the image coordinate when the vehicle exists at the grounding position y_{grnd}. In Eq. 1, since y_{grnd} is estimated from y_{grnd} itself, a feedback optimization process should be employed for the estimation.

$$y_{grnd} = y_{top} + h_{app}\left(y_{grnd}\right) \tag{1}$$

The problem now is to estimate $h_{app}(y_{grnd})$. In Eq. 2, $h_{true}(y_{grnd})$ represents the true height of the vehicle in the image coordinate, and $h_{app}(y_{grnd})$ represents appearance height of the vehicle which consists of roof length and true height of the vehicle as shown in Figure 6. Figure 6 thumbnails the side view of the road infrastructure, camera setting, and the vehicle. From Figure 6, the correction from $h_{app}(y_{grnd})$

to $h_{true}(y_{grnd})$ should be done by using correction factor $\Omega\ (y_{grnd})$ as represented in Eq. 2.

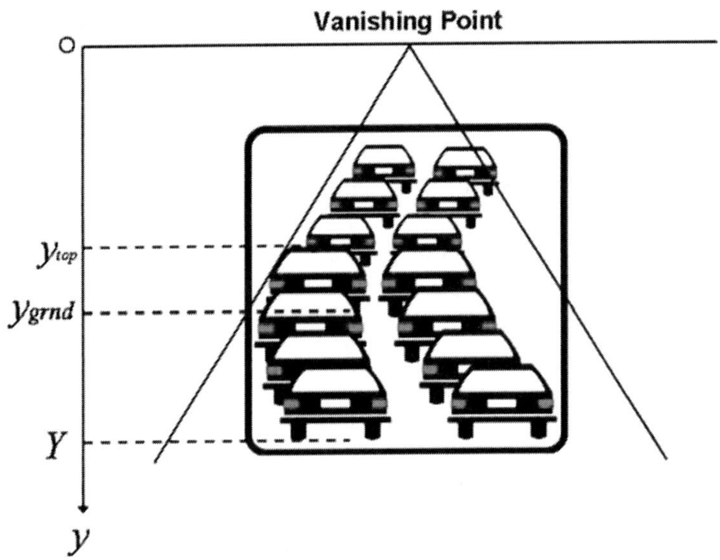

Figure 5. Definition of coordinate in the image coordinate

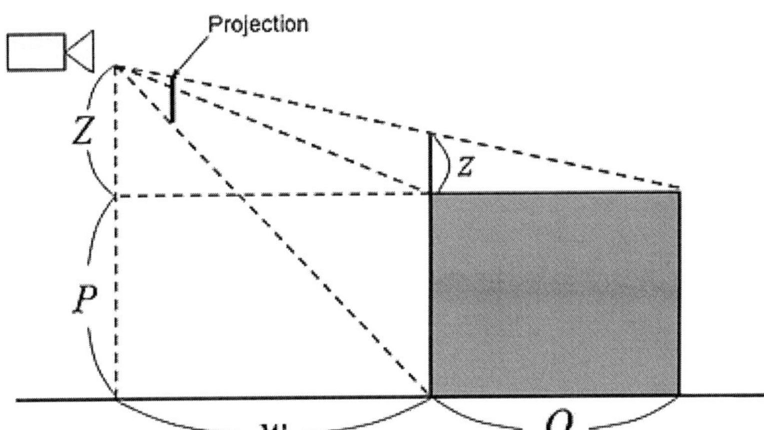

Figure 6. Side view of appearance height and true height of a vehicle projected onto the image coordinate

$$h_{app}(y_{grnd}) = h_{true}(y_{grnd}) \times \Omega(y_{grnd}) \tag{2}$$

For the purpose of the definition, Ω (*w*) is function of *w*, position in the real world coordinate, in Eq. 3. Since Ω (y_{grnd}), position in the image coordinate as a function of y_{grnd}, is used for the transformation from h_{app} (y_{grnd}) to h_{true} (y_{grnd}), calibration between *w*and y_{grnd} is required. We used cameras already in practical operations by officers in this paper. Therefore, geometrical condition of camera setup is not rigid, and it is difficult to perform calibration between *w* and y_{grnd} from the geometrical condition. In addition, since image resolution steeply declines along with the depth toward the vanishing point, it is difficult to perform the calibration accurately. Consequently, we decided to calibrate Ω (y_{grnd}) from the traffic image in direct manner as described in the experimental results, Section 6.1.

$$P \times \Omega(w) = P + z = P + Z \times \frac{Q}{w + Q} \tag{3}$$

Equation 4 assumes that true height of the vehicle in the image coordinate is proportional to the distance from the vanishing point in the image coordinate. Consequently, Eqs. 5 and 6 are obtained from Eqs. 1 to 4. As same as Eq. 2, y_{grnd} is estimated from y_{grnd} itself in Eq. 6, and a feedback optimization process is necessary. Following descriptions are about the procedure to estimate y_{grnd} of the preceding vehicle occluded by the following vehicle.

$$h_{true}(y_{grnd}) = h_{true}(Y_{grnd}) \times \frac{y_{grnd}}{Y_{grnd}} \tag{4}$$

$$\begin{aligned} h_{app}(y_{grnd}) &= h_{true}(y_{grnd}) \times \Omega(y_{grnd}) \\ &= h_{true}(Y_{grnd}) \times \frac{y_{grnd}}{Y_{grnd}} \times \Omega(y_{grnd}) \end{aligned} \tag{5}$$

$$y_{grnd} = \left\{1 - \frac{h_{app}(Y_{grnd})}{Y_{grnd}} \times \frac{\Omega(y_{grnd})}{\Omega(Y_{grnd})}\right\}^{-1} y_{top} \tag{6}$$

In the first step, h_{app} (y_{grnd}) of a vehicle is measured from the image, and the vehicle is classified into two categories of large vehicles as trucks and small vehicles as passenger cars. Vehicle regions are obtained from segmentation results from the S-T MRF model, and most of the vehicles are not occluded at the entrance area of the camera view. When the vehicle belongs to the large vehicle category, it is regarded as Ω (y_{grnd}) = 1. Since large vehicles as trucks are higher than 2.5 m, the height of the camera position, roofs of the vehicles do not appear in the image. When the vehicle belongs to the small vehicle category, Ω (y_{ground}) is obtained from the calibration results.

In the second step, y_{top} of the preceding vehicle is obtained by extracting the edge at the top of the vehicle region. The preceding vehicle and the following vehicle are segmented by the S-T MRF model against the occlusion at this moment, and the top edge of the preceding vehicle appears in the image while the bottom edge is occluded. Finally, y_{grnd} is obtained from Eq. 6 by substituting the value of y_{top}. However, right side of Eq. 6 has y_{grnd} as a valuable. Therefore, the iterative procedure should be performed as substituting y_{grnd} on the left side into y_{grnd} on the right side of Eq. 6 for the next loop of the iteration. In the first loop of the iteration, the value of y_{top} is substituted into y_{grnd} on the right side, iteration of single loop appeared as sufficient to pursue the precise result of y_{grnd} from our experiment.

Vehicle trajectory of each vehicle can be obtained by sequencing the grounding points of the vehicle in image coordinate, and the vehicle trajectory enables to estimate vehicle velocity. Thus, estimated vehicle velocity should be transformed into the velocity in the real world coordinate. Usually, calibration about the transforming parameters is performed based on solving camera parameters and setting condition. However, since the cameras used in this paper are already in practical operations aside heavy traffic, such the calibration is quite difficult. Therefore, we adopted more practical method for the calibration in this paper. Ultrasonic wave sensors are installed on the road side of our experimental field. In the highway of this paper, ultrasonic wave sensors are standardized as to show errors of less than 5% from the true value in the velocity measure. Calibrations between the image coordinates and the real world coordinates were performed to keep consistency of average velocities between the ultrasonic wave sensors and the image sensors.

5 IMPLEMENTATION OF SEMANTIC LAYER FOR INCIDENT DETECTION

Generally, it is quite difficult to detect the moment of the collision from an image. Therefore, algorithms to detect abnormally slowdown vehicles and stalling vehicles have been developed to detect incidents in which vehicles are stalling because of accidents as well as their own troubles. Slowdown vehicles and stalling vehicles are easily detected in sparse traffics. However, it is difficult to detect them from the congested or jammed traffics because a majority of vehicles move in similar velocities in congested or jammed traffics. In this paper, we designed semantic layer algorithms to classifying the behavior of each vehicle and to detect the incident by analyzing the behaviors of interacting vehicles.

5.1 Behavior-class Functions

From the previous section, we can obtain spatio-temporal trajectory of each vehicle. By classifying the trajectories, vehicle behaviors can be determined as slowdown, stalling, and lane changing. The following are examples of functions representing behaviors of vehicles such as slowdown, stalling, changing lanes, vacant space ahead of the vehicle.

- *typeOfSpTrajectory*(V.*i*, V_j,*t*): Returns 'Changing lanes' when vehicle V.*i* changes its lane to avoid vehicle V.*j*. Otherwise, returns 'Ordinary'. 'Changing lanes' includes that vehicles change its lane and has six types of traveling: 'left-to-right', 'right-to-left', 'center-to-left', 'left-to-center', 'center-to-right', and 'right-to-center'. The six types of 'Lane changing' are discriminated by evaluating similarities of a vehicle trajectory to the clustered trajectories of the particular types.

- typeOfSpTrajectory(V.i,Lkm,t)typeOfSpTrajectory(V.i,Lmk,t): Returns 'Changing lanes' when vehicle V.*i* changes its lane to avoid the m–th discretized location along the k–th lane. Otherwise, returns 'Ordinary'.

- *typeOfTpTrajectory*(V.*i*, *t*): Returns 'True' when VehicleV.*i* is stalled at time t, and 'False' otherwise. The threshold between stalled or not is determined by statistical analysis of the vehicle behaviors.

- *isSpace*(V.*i*, *t*): Returns 'True' when there is the vacant space ahead of the VehicleV.*i* along the vehicle traveling lane at time t, and 'False' otherwise. It should be noted that there is no traffic flow in any lanes, this function returns 'False'.

5.2 Event-class Functions

Event-class functions are described by combining behavior functions to determine the types of events. The following are descriptions of logics for the incident detection.

Logic 1:

If average velocity of the vehicles is less than TH_{ave_vel}, the traffic should be determined as congested, otherwise the traffic should be determined as free flow.

Logic 2:

If a vehicle is stalling at the prohibited area in the period longer than TH_{stall_time}, the scene should be determined as an incident.

Logic 3:

If there appeared the vacant space ahead of the stalling vehicle while the traffic is free flow, the scene should be determined as an incident.

Logic 4:

If there occurred lane changing behaviors over the stalling vehicle more than $TH_{lchange}$ times, the scene should be determined as an incident.

Above algorithms for the incident detection were developed and proved as successful for the incident detection at relatively high angle in the open aired highway [26]. In this paper, the algorithms are applied to the tracking results by applying the interlayer feedback algorithm to low angle images in the tunnel. We adopted $TH_{stall\_time} =$ 10 sec, and $TH_{lchange} = 3$ in the following experiments according to the publication [26], and THave_vel=20km /hourTHave_vel=20km /hour is defined according to operational rules in the highway.

6 EXPERIMENTAL RESULTS

Experiments were performed by using video images recorded by the cameras at three different locations inside the tunnel. Those three locations consist of straight, separating, and merging traffic. In this section, experimental results of vehicle segmentations and incident detections are described.

6.1 Calibration of Ω (y_{grnd})

Figure 7 shows calibration results of Ω (y_{grnd}). The plots were obtained by measuring the ratio between true height and appearance height of vehicles directly from images. All the vehicles used for the calibration were not occluded by the following vehicles, in order to obtain y_{grnd} directly from the images. Each plot corresponds to each vehicle measured. For these plots, horizontal axis represents y_{grnd} of the vehicle, and vertical axis represents Ω (y_{grnd}). Diversity of Ω (y_{grnd}) is due to variations in vehicle types, shapes of rear and top of the vehicles. Consequently, Ω (y_{grnd}) was estimated as an average of the vehicles. In Figure 7, smaller y_{grnd} means close to the vanishing point, and smaller Ω (y_{grnd}) means true height of the vehicle is close to appeared height.

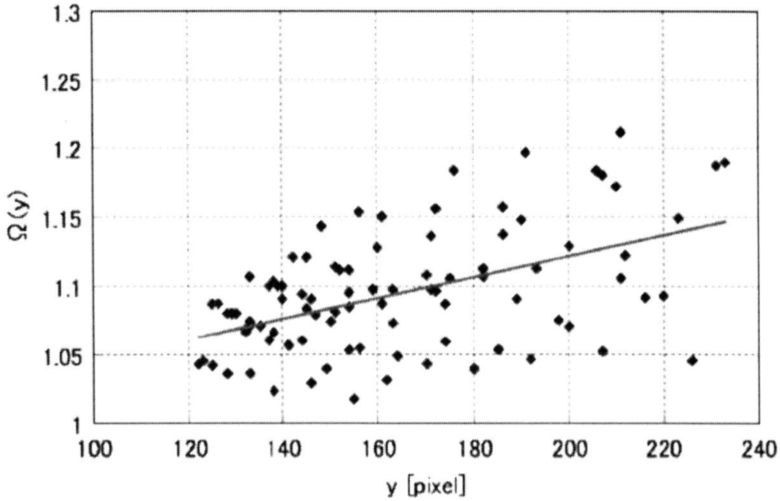

Figure 7. An example of Ω(y) calibration

6.2 Segmentation and Trajectory Estimation

Figure 8a and b exemplifies segmentation result by the S-T MRF model and the interlayer feedback algorithm. The S-T MRF model failed to segment vehicles traveling with the similar velocities and occluding along all the way in the view, as shown in Figure 8a. The interlayer feedback algorithm succeeded to segment the two vehicles, and their trajectories were correctly obtained, as shown in Figure 8b.

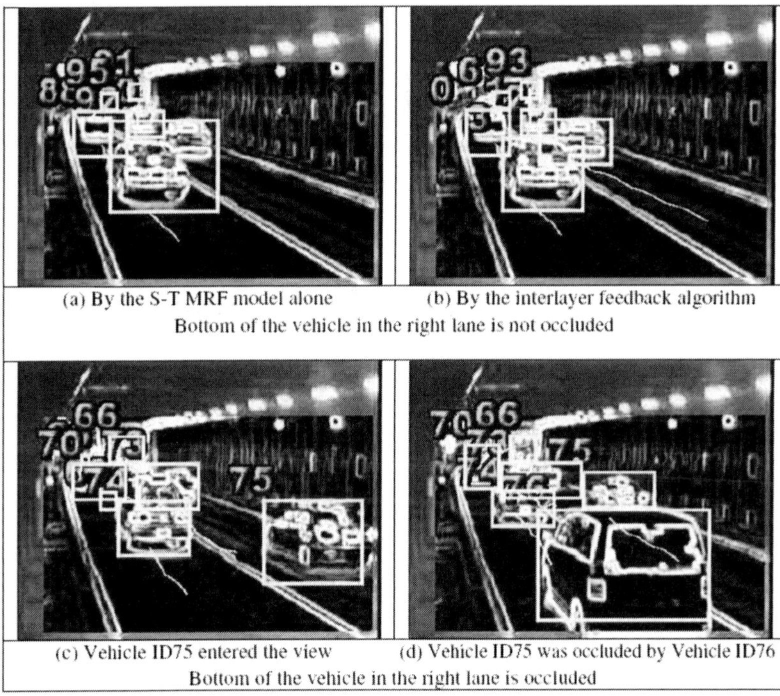

Figure 8. Segmentation results and trajectory estimations by the interlayer feedback process. **a** By the S-T MRF model alone. **b** By the interlayer feedback algorithm. Bottom of the vehicle in the right lane is not occluded. **c** Vehicle ID75 entered the view. **d** Vehicle ID75 was occluded by Vehicle ID76. Bottom of the vehicle in the right lane is occluded

Segmentation performance was evaluated by using 70-min video recorded by the three cameras in the tunnel. Totally, 8,563 vehicles were passed through the views of the three cameras. As a result, 6,577 segments were obtained by the S-T MRF model, and 8,964 segments were obtained by the interlayer feedback algorithm. Out of 8,864 segments by the interlayer algorithm, 7,856 segments were correctly appeared to include a single vehicle, and 107 segments appeared to include multiple vehicles. Nine hundred one segments appeared to be excessively divided segments. Consequently, in a rough description, contributions on vehicle segmentation against heavy occlusion would come from the S-T MRF model for about 77% and from the interlayer feedback for about 27%.

Figure 8c and d shows an example of trajectory estimations of the preceding vehicles over the occlusions by the following vehicles. Although the preceding vehicle of ID75 in the right lane was occluded by the following vehicle of ID76 in the left lane, the trajectory of the vehicle of ID75 was estimated precisely. Usually, the grounding position of the vehicle tail appears inside the bounding box of the vehicle segment. However, grounding position of the vehicle should be outside of the bounding box when the bottom part of the vehicle was occluded by the following vehicle. In Figure 8d, each line tailing from the bottom of the vehicle connects the grounding positions y_{grnd} for the past five image frames of the vehicle. The latest grounding point of the vehicle of ID75 exists outside of the bounding box. Since the bottom part of the vehicle of ID75 was occluded by the vehicle of ID76, this result was successful.

6.3 Incident Detection Results

Figure 9 shows a result of stalling vehicle detection. Figure 9a shows the scene soon after the vehicle stalled at the road side. The algorithm determined that the vehicle is stalling in the prohibited area over the emergency lane. After the length of the vehicle stalling exceeded the threshold, the state of the algorithm turned into 'incident' as in Figure 9b. Even if there was no emergency lane, the algorithm can conclude the scene as 'incident' by detecting stalling vehicle and passing vehicle in the same lane as in Figure 10a and b.

Figure 10 shows image sequences of incident detection process by the S-T MRF model alone and the interlayer feedback algorithm. The incident could not be detected by the S-T MRF model alone, since vehicles could not be segmented because of heavy occlusions as shown in Figure 10a. However, the incident was detected successfully by the interlayer feedback algorithm and edge pattern classification, since vehicles were segmented precisely against heavy occlusions as shown in Figure 10b. In this case, the algorithm detected the incident by detecting frequent lane changing behaviors against a stalling vehicle. Such the behavior analyses over the congested traffic were enabled by the vehicle segmentation using the interlayer algorithm and the trajectory estimations against occlusions.

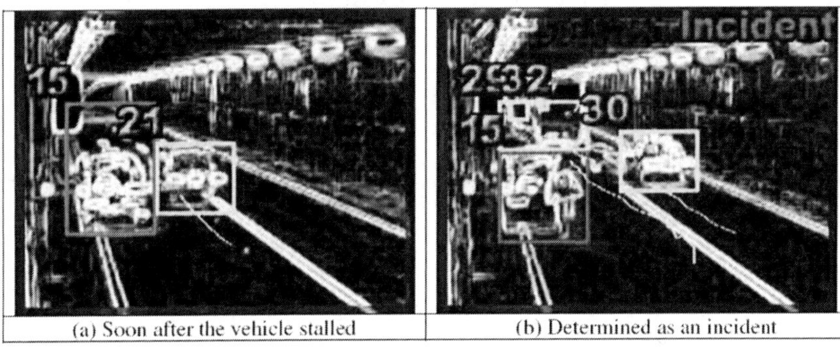

Figure 9. Detection of stalling vehicle. **a** Soon after the vehicle stalled. **b** Determined as an incident

Consequently, experimental results were summarized as shown in Table 1, in terms of success rates of incident detection. For the experiments, 32 incidents were edited from the 6 month video stream of the cameras. Videos of 16 incidents were used as learning data to determine thresholds to detect slowdowns, stalling, and the frequent lane changing in the same location to avoid the incident vehicles. The other videos of 16 incidents were used as test data to examine the performance of the algorithm. Success rates for incident detections were improved by the interlayer feedback algorithm to 88% for test data and 91% in total of learning data and test data, whereas they were 44% and 53% by the S-T MRF model alone, respectively.

Table 1. Incident detection results: success rates

	Learning Data	Test Data	Total
By the S-T MRF model alone	10/16=63%	7/16=44%	17/32=53%
By the interlayer feedback	15/16=94%	14/16=88%	29/32=91%

(a) By the S-T MRF model | (b) By the interlayer feedback algorithm

Figure 10. A result of Incident detection: lane changing around the stalling vehicle

7 CONCLUSIONS

In this paper, an algorithm for incident detection in the tunnel was developed. We developed the interlayer algorithm to combine the S-T MRF model and the algorithm in the Morphology layer, in order to resolve heavy occlusion problems in heavy traffic with the low angle image. The algorithm was proved to be effective for the segmentation of vehicles in the heavy occlusion images though the examinations using heavy occlusion images. Consequently, the algorithm could obtained precise trajectories of the vehicles, and contributed to the incident detections from the congested traffics in tunnels.

Since the incidents in tunnels may cause additional accidents or fires that may result in fatal disaster, it is important to detect them as soon as possible. The disaster would be rather serious in the heavy traffics of urban tunnels. The technology in this paper is expected to enhance the safety of tunnel traffics and the efficiency of tunnel operations.

REFERENCES

1. Kolling, H., Nagel, H.: 3D pose estimation by directly matching polyhedral models to gray value gradients. IJCV **23**(3), 283–302 (1997)
2. Peterfreund, N.: Robust tracking of position and velocity with Kalman Snakes. IEEE Trans. Pattern Analysis and Machine Intelligence (PAMI), vol. 21, no. 6, pp. 321–331, 1988
3. Kass, M., Witkin, A., Terzopoulos, D.: Snakes: active contour models. Int. J. Comput. Vis. **1**, 564–569 (1999)
4. Smith, S.M., Brady, J.M.: ASSET-2: real-time motion segmentation and shape tracking. IEEE Trans. Pattern Analysis and Machine Intelligence (PAMI), vol. 17, no. 8, pp. 814–820, 1995
5. Stauffer, C., Grimson, W.E.L.: Adaptive background mixture models for real-time tracking. Proc. of CVPR, 1999, vol. II, pp. 246–252, Jun 1999
6. Leuck, H., Nagel, H.: Automatic differentiation facilitates of-integration into steering-angle-based road vehicle tracking. IEEE CVPR'99, pp. 360–365

7. Gardner, W.F., Lawton, D.T.: Interactive model-based vehicle tracking. IEEE Trans. PAMI, vol. 18, no. 11, pp. 1115–1121, 1996

8. Weber, K., Malik, J.: Robust multiple car tracking with occlusion reasoning. Proc. of ECCV'94, Vol. I, Lecture Notes in Computer Science 800, pp. 189–196. Springer-Verlag

9. Hsieh, J.-W., Yu, S.-H., Chen, Y.-S., Hu, W.-F.: Automatic traffic surveillance system for vehicle tracking and classification. IEEE Transactions On Intelligent Transportation Systems, vol. 7, no. 2, pp. 175–187, June 2006

10. Kanhere, N.K., Birchfield, S.T.: Real-time incremental segmentation and tracking of vehicles at low camera angles using stable features. IEEE Transactions On Intelligent Transportation Systems, vol. 9, no. 1, pp. 148–160, March 2008

11. Wang, C.-C.R., Lien, J.-J.J.: Automatic vehicle detection using local features - a statistical approach. IEEE Transactions On Intelligent Transportation Systems, vol. 9, no. 1, pp. 83–96, March 2008

12. Zhang, W., Wu, Q.M.J., Yang, X., Fang, X.: Multilever framework to detect and handle vehicle occlusion. IEEE Transactions on Intelligent Transportation Systems, vol. 9, no. 1, pp. 161–174, March 2008

13. Ambai, M., Ozawa, S.: Vision-based traffic monitoring system for various traffic situations using monocular camera. 13th World Congress on ITS, London, October 2006

14. Cucchara, R., Piccardi, M., Mello, P.: Image analysis and rule-base reasoning for a traffic monitoring system. IEEE/IEEJ/JSAI ITS conf., Tokyo, pp. 758–763, Oct. 1999

15. Jung, Y., Lee, K., Ho, Y.: Content-based event retrieval using semantic scene interpretation for automated traffic surveillance. IEEE Trans. on ITS, Sep. 2001

16. Zhang, K., Taylor, M.: A new method for incident detection on urban arterial roads. 11thWorld Congresson ITS, Nagoya, Oct 2004

17. Srinivasan, D., Jin, X., Cheu, R.L.: Evaluation of adaptive neural network models for freeway incident detection. IEEE Trans. ITS, vol. 5, no. 1, March 2004

18. Byun, S.C., Choi, D.B., Ahn, B.H., Ko, H.: Traffic incident detection using evidential reasoning based data fusion. Proceedings of ITS 1999, 6th World Congress on ITS

19. Thomas, T., van Berkum, E.C.: Detection of incidents and events in urban networks. IET Intell. Transp. Syst.**3**(2), 198–205 (2009)

20. Zheng, P., McDonald, M., Jeffery, D.: Event detection based on loop and journey time data. IET Intell. Transp. Syst. **2**(2), 113–119 (2008)

21. Shehata, M.S., Cai, J., Badawy, W.M., Burr, T.W., Pervez, M.S., Johannesson, R.J., Radmanesh, A.: Video-based automatic incident detection for smart roads: the outdoor environmental challenges regarding false alarms. IEEE Transactions on Intelligent Transportation Systems, vol. 9, no. 2, pp. 349–359, June 2008

22. Versavel, J.: Improving road and tunnel safety via incident management: implementing a video image processing system. 16th World Congress on ITS, Stockholm, September 2009

23. Austin, M., Brock, M., Carter, E., Metcalfe, F., Orme, L.: Implementation and evaluation of an adaptive CCTV display system for incident detection for the highways agency. 16th World Congress on ITS, Stockholm, September 2009

24. Kamijo, S., Matsushita, Y., Ikeuchi, K., Sakauchi, M.: Traffic monitoring and accident detection at intersections. IEEE trans. ITS, vol. 1, no. 2, pp. 108–118, June 2000

25. Kamijo, S., Nishida, T., Sakauchi, M.: Occlusion robust and illumination invariant vehicle tracking for acquiring detailed statistics from traffic images. To appear in IEICE Transactions on Information and Systems, vol. E85-D, no. 11, 2002

26. Kamijo, S., Harada, M., Sakauchi, M.: An incident detection system based on semantic hierarchy. Proceedings of ITSC 2004, pp. 853–858, Oct. 2004

Index

N

O

P

Q

R

S

T

U

V